M000303629

ZOOLOGY

OF

THE VOYAGE OF H.M.S. BEAGLE,

UNDER THE COMMAND OF CAPTAIN FITZROY, R.N.,

DURING THE YEARS

1832 TO 1836.

*PUBLISHED WITH THE APPROVAL OF
THE LORDS COMMISSIONERS OF HER MAJESTY'S TREASURY.*

Edited and Superintended by

CHARLES DARWIN, ESQ. M.A., F.R.S., V.P.G.S.,

NATURALIST TO THE EXPEDITION.

PART V.

REPTILES,

BY

THOMAS BELL, ESQ., F.R.S., F.L.S., &c.

PROFESSOR OF ZOOLOGY IN KING'S COLLEGE.

British Library Cataloguing-in-Publication Data
A catalogue record for this book is available from
the British Library

PREFACE.

AMONGST the Reptiles and Amphibians obtained by Mr. Darwin, in the Voyage of the Beagle, there are several of great interest, not merely on account of their novelty as newly discovered species, of which there are nearly thirty, or as forming the types of genera not previously known, or of any remarkable peculiarity of form, structure, or habit, although in all these respects many of them are highly interesting; but more particularly as serving to establish or confirm several points connected with their geographical distribution.

From the structure of most of these animals and their consequent habits of life, circumscribed as they are for the most part in their locomotive powers, it might reasonably be predicated that they would, upon the whole, exhibit as distinct examples of restriction, with regard to their geographical boundaries, as any class of vertebrated animals; and that the intervention of seas and of mountains would be sufficient to limit the range of a species. Such is in fact usually the case; and not only is the same species not found in the Old and New Continents, but, with very few exceptions, not even on the opposite sides of the South American Continent, in which range Mr. Darwin's discoveries have principally been made. The occurrence, however, of *Bufo Chilensis* at Rio Janeiro and at Buenos Ayres on the eastern, and at Valparaiso and the Archipelago of Chonos on the western side of the continent, shows an extent of distribution exceedingly unusual if not absolutely unparalleled in this family. It is, however, still possible that further and more extended researches into the characters of the animals in question, and an examination of individuals from each locality at various ages, may prove that there are two species, which have been confounded with each other, and the anomaly may thus be removed.

But although the circumscribed range of a species may be accounted for by the reasons above mentioned, and others of a restrictive nature, it is not so easy to refer to any known or obvious cause the remarkable fact of a whole genus, consisting of numerous species, being thus geographically limited. Yet this is a well-known and very common circumstance with regard to several groups of animals. In our present researches there exists a remarkable example of this fact in the genus *Proctotretus,* consisting, as is now known, of at least fourteen species, all inhabiting the western coast of South America. These facts, interesting as they are, have never been sufficiently investigated, although, it must be confessed, there are so many anomalies in relation to this subject, that we must despair of ever reducing the facts in question to any thing like fixed laws.

The close approximation of the Raniform and Hyliform groups of the Anourous Amphibia is strikingly illustrated by several new forms obtained by Mr. Darwin, which are so perfectly osculant between the two families, that it is difficult to assign them a decided location. And the addition of some bufonine forms in the family *Ranidæ,* as at present constituted, and on the other hand of some amongst the *Bufonidæ,* which are no less raniform in their general structure and habits, render it increasingly probable that the single character of the presence or absence of superior maxillary teeth, must be considered as in-sufficient to constitute alone a natural distinctive family character. There are several minor points bearing upon the natural arrangement of the Anourous Amphibians, which are illustrated by the characters of some of the species now first described, which will doubtless at some future time assist in the construction of a classification of these animals, bearing at least a nearer approximation to their natural arrangement than any that has hitherto been promulgated.

The Ophidians have been placed in the hands of Mons. Bibron, who is at the present time engaged in completing his admirable history of Reptiles, by the publication of those volumes which are devoted to this order; and it must be considered a fortunate circumstance that the delay which has taken place in the appearance of that portion of his labours, has thus afforded the opportunity of embodying in so perfect a work, the numerous discoveries of Mr. Darwin in this particular department of Erpetology.

                                                                    T. B.

Hornsey, Sept. 2, 1843.

LIST OF SPECIES.

LIST OF PLATES.

Errata.—In Plate XIX. *for* "Hylonia" *read* "Hylorina."
for "vanterii" *read* "Vauterii."

REPTILES.

TRIBUS—EUNOTES. *Bibron.*

FAMILIA—TROPIDURIDÆ. *Mihi.*

TROPIDOLEPIDIENS. *Bibr.*

GENUS—Tropidurus. (In part.) *Weigmann.*

PROCTOTRETUS. *Bibr.*

CHARACTER GENERICUS.—*Pori femorales* nulli. *Pori præ-anales* in maribus tantum. *Crista dorsalis* nulla. *Dentes palatini. Squamæ* imbricatæ; *dorsales* carinatæ.

THE genus which I take the present favourable opportunity to illustrate, formed a section or sub-genus of the genus *Tropidurus,* according to Weigmann, who, how-ever, was acquainted with two species only; *Pr. Chilensis* and *Pr. nigromaculatus.* Of the varieties of the former of these, that author has made no less than three species; but these have been very properly reduced by M. Bibron to one only. The last-named excellent erpetologist described several additional species, which had been brought from Chile, by M. D'Orbigny, and others; and I received, some years since, from Capt. King, three or four species which were found by him in the same locality, in the course of his well-known survey. The number of species altogether, hitherto known, amounts to ten; to which I have now the opportunity of adding four entirely new, forming part of the interesting collection of Reptilia made by Mr. Darwin. One of them, *Pr. Kingii,* was already in my collection, amongst those which were given me by Capt. King. The genus, therefore, of which, but lately, two species only were known, now consists of fourteen; and it is highly probable that more may yet be obtained by more pro-longed and extensive investigation in the same districts.

B

Thus of the species now described two were known to Weigmann, and described by him, namely, *Chilensis*, and *nigromaculatus*. I received from Capt. King, *Chilensis*, *pictus*, *cyanogaster*, *Kingii*, *Fitzingerii*, and, from other sources, *Chilensis*, *pictus*, *Weigmannii*. Bibron describes the whole of these, excepting *Kingii*, and in addition to them *tenuis*, *pectinatus*, *signifer*. In Mr. Darwin's collection are found all those described by Bibron, excepting *signifer*; and in addition *Kingii* now first described, and the following species never before observed, namely, *Darwinii, gracilis, Bibronii.*

Although the form of the whole of the species much more nearly approximates that of the Agamidæ than most others, they are far removed from that family by several important characters, which it is unnecessary here to detail.

Proctotretus Chilensis.

Plate I.—Fig. 1.

Auribus margine anteriore dentato; collo non plicato; squamis dorsi magnis, rhombeis; acutè carinatis; serie unicá squamarum supralabialium.

Calotes *Chilensis,* Less. et Garn. Voy. de la Coquille, Zool. Rep. t. i. f. 2.
Tropidurus *Chilensis,* Weigm. Act. Acad. Cæs. Leop. Carol. Nat. cur. xvii. pp. 233. 268.
Proctotretus *Chilensis,* Bibr. Hist. Nat. des Rep. IV. p. 269.
Sun. Tropidurus *nitidus,* Weigm. l. c. p. 234, t. xvii. f. 2.
Var. Tr. *olivaceus,* Ib. l. c. p. 268.

Habitat, Guasco in Northern Chile.

Description.—Head short and broad, rostrum rounded, obtuse. Scales of the head large, and slightly raised, separated by distinct grooves. Superciliary ridge strongly marked, forming a distinct carina, composed of five or six narrow, elongated, obliquely imbricated scales. Nostrils large, nearly round. A single series of narrow scales between those of the upper lip and the orbit. Scales of the temples imbricated, rhomboidal and carinated. The opening of the ear oval, rather large, furnished anteriorly with three or four projecting scales, of which the upper one is the largest. The neck is short, robust and round, and without any lateral fold; in which it differs from every other species of the genus. The trunk is thick, rounded on the back and sides, flattened beneath, diminishing toward either extremity. The tail is ordinarily almost twice as long as the body, thick at its origin, and tapering regularly to the extremity, nearly round, excepting near the base, where it is slightly quadrilateral. The anterior feet when placed against the sides, extend backwards little more than mid-way between the shoulder and the groin; the posterior ones, stretched forward, reach the arm-pit.

The scales of the whole upper and lateral parts of the body, tail, and limbs, are loosely imbricated, large, rhomboidal, and furnished with an elevated carina, terminating in an acute point: those of the under parts are large, smooth, and obtuse.

This is one of the most beautiful species of the genus. The general form is robust and solid, forming a remarkable contrast with most other species of the genus. The surface is beautifully relieved by the fine, large and prominent scales, which are ranged in rows of perfect regularity, of which there are about eighteen on the back and side.

In colour and markings the individuals differ considerably; so as to have given rise to the opinion that they form three distinct species. Monsieur Bibron mentions two principal varieties, constituting the *Tropidurus olivaceus* and *Tr. Chilensis* of Weigmann. In the former, the colour is of a more or less bronzed bright green, or yellowish, according as the green or yellow colour prevails on the scales, each of which is green, with a border of yellow on each side : this border, according to M. Bibron, in some individuals of a blood-red colour. In some, especially in young individuals, there are several waved bands running transversely across the back.

In the second variety, says M. Bibron, the upper parts are either of an olive colour, with a golden glance in certain lights, or fulvous, with more or less of a yellow tint; and these have always four brown bands running the length of the body, appearing as if formed of a series of large spots united together. " The temples are marked with a black line, which extends from the posterior angle of the eye to the ear; another divides behind the occiput into two branches, which pass to the neck there to join the dorsal bands." There are other variations of colour, but scarcely deserving to be considered as constituting permanent or fixed varieties; the largest and finest specimen I have seen, which I received from Chile, is almost uniformly of a fine metallic green, without any markings.

DIMENSIONS.

	Inches.	Lines.
Length of the head	0	9
of the body	3	5
of the tail	6	0
Total length	10	4
Length of anterior extremity	2	0
of posterior extremity	1	2

This species would appear to be very common in Chile, from the numerous specimens from that country in the museum in Paris, which were brought by M. Gay, and by M. D'Orbigny. I have also received specimens from Capt. King. I find only one or two specimens in Mr. Darwin's collection, which he found at Guasco in Chile.

PROCTOTRETUS GRACILIS. N.S.

PLATE I.—FIG. 2.

Corpore gracili: capitis squamis lævibus, non imbricatis: aurium margine anteriore minutè bi-tridentato: collo vix plicato, squamis imbricatis: serie unicâ squamarun supralabialium: femorum facie posteriore omninò granulosâ.

Habitat, Port Desire, Patagonia.　Mr. Darwin, MS.

DESCRIPTION.—This new species is more slender and graceful in its general form than any other of the genus, not excepting *Pr. tenuis*, which in its general proportions it considerably resembles. The head is rather short, the anterior portion including the eyes being nearly an equilateral triangle. The muzzle is rounded. The scales of the head flat, rather large, consisting behind the nose of three series of 2, 3, 3, and 2, mostly hexagons. The nostrils are small and perfectly round, the superciliary ridge is very slightly marked; there is but a single row of small linear scales between the labial and suborbitar. The ear is of moderate size, the anterior margin having two or three small projecting scales. The scales on the temples are imbricated and smooth; those of the sides of the neck also imbricated but very small. The neck is almost wholly without a fold, having merely a slight loose elevation of the skin. The scales of the back are small, rhomboidal, flat, the carina low, and not pointed at the extremity. They consist of about ten rows on each side of the median line. Those of the sides and belly are wholly without any notch on the free margin. The scales around the axillæ, and those of the posterior face of the thighs are granular. The tail is of considerable length, being more than twice the length of the head and body. The limbs are remarkably long; the anterior, when pressed against the side, reaching to the setting on of the thigh, and the posterior reaching forwards nearly to the ear. The general colour of the upper parts is greyish brown, with a yellow longitudinal fascia extending on each side from the upper edge of the orbit to some distance along the tail—and another from beneath the eye to the thigh. The middle of the back is lighter than the sides—beneath the second lateral line the colour fades, and on the belly it is pale buff or light yellow. The sides are dotted with black; and there are some dark waved lines and dots beneath the lower jaw.

　　This species is at once distinguished from every other by the fold of the neck being scarcely cognizable. It is however not so absolutely wanting as in *Pr. Chilensis.*

DIMENSIONS.

	Inches.	Lines.
Length of the head........................	0	5
of the body	1	4
of the tail	3	8
Total length......	5	7
Length of anterior extremity	0	7
of posterior extremity	1	1

A single individual of this species was taken by Mr. Darwin at Port Desire.

5

PROCTOTRETUS PICTUS.

PLATE II.—FIG. 1, 2.

Capite squamis parvis, lævibus, non imbricatis ; aurium margine anteriore granuloso;
serie unicâ squamarum supralabialium ; squamis temporum subcarinatis, imbricatis ;
collo ad latera granuloso; squamis dorsalibus parvis, rhombeis, carinâ humili, posticè
obtusâ ; facie posteriore femorum omninò granulosâ.

Proctotretus pictus, Bibr. l. c. p. 276.

Habitat, Chile.

DESCRIPTION.—This species is moderately slender in its general form, but more fusiform than *Pr. tenuis,* which it much resembles in many of its characters. The head, which is rather short, and with the snout rounded, is covered with numerous small flat plates which vary exceedingly in their form and arrangement. The upper surface of the head is flattened, and the superciliary ridges distinctly marked. The temples are covered with small flat scales, which are slightly imbricated and carinated. The sides of the neck are granulated; and the anterior margin of the auditory cavity has small simple granulations. The scales of the back are distinguished from those of many other species by the flatness of the carina which is also obtuse posteriorly; they are small and closely imbricated. Those of the sides are almost without any carina, and those of the belly and throat small and very smooth, and the whole of them entire. The upper parts of the limbs are covered with scales similar to those of the back, but smaller. Those of the under part of the fore arm are similar, but beneath the thighs they are smooth and on the posterior part of the thighs they are wholly granular. The tail is furnished with quadrilateral carinated scales disposed in whorls. The margin of the cloaca has from two to four pores.

The colour of this species varies greatly. Bibron has enumerated three principal varieties, of which I have several specimens, which were brought home and presented to me by Capt. King, who obtained them during his survey of the coast of South America. These varieties, however, occasionally run into each other.

Var. A. General colour of the upper part bronzed or coppery, having a green longitudinal line on each side of the back, at the inner margin of which is a series of very distinct black dots. The sides of the neck and body are of a similar colour to the back, with indistinct black spots ; beneath this part the ground colour becomes blue with black dots. The throat is blackish, and the inferior surface generally is very pale bluish green.

Var. B. This variety is described by M. Bibron as of a brown colour more or less dotted with yellow, and having a line of that colour along each side of the back, extending from the posterior angle of the eye to the base of the tail, and having on each side a series of angular black spots. Some of these spots on the upper part of the flanks, become dilated, so as to form a sort of vertical or transverse waved bands, with yellowish margins. On the neck there are small black lines, and the upper part of the head is brown with blackish spots. The upper surface of the legs and of the tail is brown with transverse bands composed of black dots. The whole under surface is of a whitish colour, sometimes having a slight tint of orange towards the posterior parts, marbled with black.

Var. C. In this variety the general colour is dark brown, and the yellow or green longitudinal lines which characterize the former varieties are but slightly marked ; but the black spots unite and form irregular transverse bands.

DIMENSIONS.

	Inches.	Lines.
Length of the head	0	6
of the body	1	8
of the tail	4	6
	7	0

This specimen very much resembles in its more tangible characters, the *Pr. tenuis;* from which, however, it differs totally in the colouring, and in some measure also in the general form, which is more thick and robust.

Found in Chile by M. Gay, from whence I also received specimens from Capt. King, and other sources.—" Valparaiso." Mr. Darwin.

PROCTOTRETUS BIBRONII. N.S.

PLATE III.—FIG. 1.

Capite squamis lævibus, subconvexis ; auribus ovalibus, margine anteriore unidentato ; squamis temporum collique rotundatis lævibus imbricatis ; colli minimis ; serie unicâ squamarum supralabialium ; squamis dorsi rhomboideis, carinatis, posticè acuminatis ; abdominis squamis omnibus integris ; femorum fucie posteriori omnino granulosâ.

Habitat, Port Desire. Mr. Darwin.

DESCRIPTION.—General form resembling that of *Pr. pictus* and *cyanogaster.* Head moderately short, obtuse, covered with rather large slightly convex scales; a single row of scales between the labial and the orbitar. The anterior margin of the ear has a single tooth. The temples and the sides of the neck are covered with imbricated scales, which have no carina—those of the neck, especially those on the fold of the skin are smaller and more raised than the others. The scales of the back are rather large, rhomboidal, with a distinct carina, terminating in a point. Those of the abdomen and sides are all of them entire at the margin. The posterior surface of the thighs is wholly granular.

The only specimen obtained being a female, the number of pre-anal pores is not known.

The general colour of this species is brownish grey ; a black longitudinal line runs down the middle of the back and tail. There are two series of black spots on each side, and a small interrupted fascia of the same colour extends from the shoulder to the thigh. The belly is of an uniform dirty white.

This species approaches considerably to *Pr. cyanogaster* in general form and habit, and in many of its characters; but it may be at once distinguished from it not only by its colouring, but by the absence of even the slightest appearance of a carina on any of the scales of the temples or of the sides of the neck.

DIMENSIONS.

	Inches.	Lines.
Length of the head......................	0	6
of the body	1	6
of the tail	3	4
Total length......	5	6
Length of anterior extremity	0	7
of posterior extremity	1	1

Found by Mr. Darwin at Port Desire, in Patagonia.

PROCTOTRETUS TENUIS.

PLATE III.—FIG. 2.

Capite squamis lævibus, non imbricatis; auribus magnis, margine anteriore sub-tuberculato; serie unicâ squamarum supralabialium; temporibus squamis rotun-datis, imbricatis; collo granuloso; squamis dorsi parvis, obtusis, carinis minimis; squamis lateralibus exiguis, non imbricatis; facie posteriore femorum omninò granulosâ.

Proctotretus tenuis, Bibr. l. c. p. 279.

DESCRIPTION.—General form slender: head rather short and obtuse, covered with flattened smooth scales; anterior margin of the ears with one or more slight tubercles; temples covered with rounded imbricated scales, some of which are slightly carinated; sides of the neck, and above the shoulders granular; scales of the back small, slightly carinated, obtuse; those of the sides very small, very little imbricated; those of the belly small and smooth. The posterior surface of the thighs has no patch of imbricated scales, but is wholly granular.

The colour of the two specimens brought home by Mr. Darwin is so much obliterated, that I am obliged to have recourse to the account given by Bibron of the colour and markings of this species: — "Les deux sexes du *Proctotréte svelte* n'ont pas le même mode de coloration. Ni l'un ni l'autre ne portent, de chaque côté du dos, une bande longitudinale verte ou jaunâtre comme cela s'observe dans l'espèce précédente, (*Pr. pictus.*)

" Le mâle a le dessus de la tête nuancé de brun et de fauve, ou bien ponctué de jaune et de noirâtre. La région cervicale est, ainsi que le dos, vermiculéc de noir sur un fond brun, qui est lui-même semé de tachcs, soit bleuâtres, soit verdâtres, on ardoisées ; quelquefois même on en remarquc de jaunâtres. Presque tous les individus ont les côtés du eou marqués chaqun d'une raie noire qui s'étend depuis le haut dc l'oreille jusqu'a l'épaule. Les membres et la queue sont coupés en travers par des bandes onduleuses noirâtres, dont les intervalles se trouvent remplis par de taches, les unes bleuâtres, les autres de la couleur du euivre rouge, la gorge tantot est jaune, tantot d'un beau vert métallique. Souvent clle est, de même que les autres régions inferieures de l'animal, vermieulée de gris-brun pâle sur un fond blanchâtres, glacé dc violet.

"La femelle a toutcs scs parties supérieurcs peintes d'un gris-brun fauve. Son cou et son dos portcnt deux séries parallèlc de demi-ecreles noirs, ayant leur bord convexe tourné de eoté de la tête, et leur bord coneave liseré de blanehâtre, ou bien d'une teintc plus claire que celle du fond de la couleur du dos. La région moyenne de celuici est quelquefois ponetuée de noir, ou tachetée de blanehâtre. Des lignes noires onduleuses traversent le dessus de la queue dont le dessous est souvent euivreux. Les régions inférieures sont blanchâtres on bien colorées de la mêmc manière que celles des individus mâles."

Mr. Darwin's only observation on the eolour of this specics is, that it is "brownish black with transverse blaek bands."

DIMENSIONS.

	Inches.	Lines.
Length of the head........................	0	5
of the body	1	6
of the tail	2	1
Total length......	4	2
Length of anterior extremity	0	8
of posterior extremity...........	1	3

Found at Valparaiso, and at Concepcion, in Chile.

PROCTOTRETUS SIGNIFER.

PLATE IV.—FIG. 1.

Capite brevi, obtuso, depresso, squamis lævibus planis; aurium margine anteriore bi-tuberculato ; squamis temporum imbricatis; colli granulatis ; serie unicâ squamarum supralabialium ; squamis dorsi laxis imbricatis, vix carinatis ; facie posteriore femorum omninò granulosa. Dorso flavescenti-griseo, signis nigris, in seriebus quatuor longitudinalibus dispositis.

Proctotretus signifer, Bibr. l. c. p. 288.

Head short, dcpressed, somewhat abruptly deflcxed from the vertex—the seales flat, those of the supra-orbital arch being numerous, and less regular than in most other species. Between the labial scales, and the long infra-orbital plate is a single series of rounded scales.* The scales of the temples are rather large, somewhat rounded, slightly imbricated, and a few of the posterior ones having the vestige of an obtuse carina. The ear is rather small, the anterior margin having two slightly prominent scales towards the lower part.

The body is depressed; the tail moderately long, thick and slightly four-sided at the base, becoming much smaller and round towards the middle. The scales of the sides of the neck are small and granular; those of the upper parts of the body small, rhomboidal, rounded posteriorly, loose, much imbricated, and with an extremely low and inconspicuous carina. They are arranged in about twenty-two longitudinal series. Those of the sides are larger and quite smooth. The scales of the whole under part of the throat and belly are rhomboidal, smooth and much imbricated; a very few towards the sides of the abdomen are slightly notched at the apex. The under surface of the anterior and the hinder surface of the posterior extremities are covered with very fine granular scales; those of the upper surface of the members being rhomboidal, smooth, but slightly carinated and obtuse. The tail is covered with small rhomboidal scales which are considerably imbricated and distinctly carinated. The scales of the upper surface of the toes are smooth; those on their sides uni-carinated, and those beneath tri-carinated and broad.

The general colour of the upper parts is a yellowish grey, with black markings, which have somewhat the character, as Mons. Bibron observes, of Arabic letters. On the neck and back these markings are disposed in four longitudinal series; and there are small linear markings on the upper part of the shoulders and thighs. The tail is similarly marked, the under parts are whitish, with brown lines and spots.

DIMENSIONS.

	Inches.	Lines.
Length of the head	0	6
of the body	1	5
of the tail	3	0
Total length	5	1
Length of anterior extremity	0	8
of posterior extremity	1	2

This species is not found amongst the reptiles obtained by Mr. Darwin, but as it has never been figured, it appeared very desirable that this opportunity should not be lost. I am indebted to the great kindness of my friend M. Bibron for the loan of the only specimen which I have seen, and from which the accompanying figure is taken. It formed part of the zoological collections obtained by Mons. D'Orbigny for the French Museum.

* Mons. Bibron states that there are two series, but on examining his specimen I find a single series only.

C

PROCTOTRETUS NIGROMACULATUS.

PLATE IV.—FIG. 2.

Capite brevi, squamis neque imbricatis nec carinatis, tecto ; auribus margine anteriore tridentato ; serie unicâ squamarum supralabialium ; squamis temporum magnis, rotundatis, imbricatis ; colli (et præcipuè plicæ) crassis, elevatis ; facie posteriore femorum omninò granulosâ ; squamis nonnullis ad latera abdominis gulæque emarginatis ; maculâ transversè oblongâ, nigrâ, supra regionem scapularem.

Tropidurus (Leiolæmus) nigromaculatus, Weigm. Act. Acad. Cæs. Nat. cur. xvii. p. 229.
Proctotretus nigromaculatus, Bibr. l. c. p. 281.

Habitat, Coquimbo, Chile.

DESCRIPTION.—Head short, the muzzle rounded, rather obtuse ; scales of the upper part of the head somewhat convex, smooth ; a single line of scales between the labial and orbitar ; ears rather large, the anterior margin having three rather prominent scales, the middle one being the largest. Temporal scales somewhat large, smooth, rounded, and imbricated, those towards the upper part slightly carinated ; those of the sides of the neck thick and elevated, those on the fold being rather acutely prominent ; scales of the back carinated, the carina terminating in an acute point ; a few of the scales at the sides of the abdomen and throat notched ; the remainder of the scales of the under parts rhomboidal. The scales about the axilla and around the base of the shoulder are granular and very small, as are those of the posterior surface of the thighs, on which part there are no imbricated scales as in *Tr. Fitzingerii.* The tail is round, excepting at the base where it is slightly flattened ; it is moderately long and slender. The anterior extremity placed against the side does not reach to the thigh ; the posterior similarly placed reaches to the shoulder. The colour of this species is yellowish grey above, with a yellowish longitudinal line on each side the back, and two rows of black spots each margined with yellow behind. There is a large and distinct oblong black mark on the region of the scapula, from which circumstance the species takes its name. On the posterior surface of the thighs, towards the upper part, are three or four black dots placed in a line. The under surface is yellowish white with dark grey dots and lines under the chin and throat.

DIMENSIONS.

	Inches.	Lines.
Length of the head...............	0	5
of the body	1	5
of the tail	2	8
Total length....	4	8
Length of anterior extremity	0	8
of posterior extremity	1	2

This species was first described by Weigmann, and subsequently by M. Bibron, from specimens obtained by Gaudichaud from Coquimbo, at which place the single specimen brought home by Mr. Darwin was also obtained.

PROCTOTRETUS FITZINGERII.

PLATE V.—FIG. 1.

Capite squamis lævibus, non imbricatis; margine anteriore aurium granuloso; squamis supralabialibus ovalibus, in serie unicâ dispositis; squamis dorsalibus parum carinatis, postice obtusis. Facie posteriore femorum præcipuè granulosâ sed portione, caudam versus, squamis majoribus, rhomboideis imbricatis tectâ.

Proctotretus Fitzingerii, Bibr. l. c. p. 286.

Habitat, Patagonia.

DESCRIPTION.—General form thick and robust, the head short being nearly as broad as it is long. The muzzle slightly rounded. Scales of the head flat, small and numerous. Two scales only behind the rostral and between those which are pierced by the nostrils. Those over the nose and around the occipital scales being larger and more regularly arranged than the others. The ear is large, oval, the anterior margin having, towards the upper part, about three small, oval, granular, very slightly projecting scales. Temples covered with small, rounded or slightly hexagonal scales, which are scarcely imbricated. A single range of rather broad oval scales between the orbit and the upper lip. Scales of the sides of the neck, and above and behind the shoulder small, granular. The trunk thick; scales of the back very small, imbricated, very slightly carinated, and not pointed behind; those of the under parts smooth and rhomboidal. The legs are short and robust. The anterior pair, placed against the side do not extend backwards more than halfway to the thigh. The posterior pair do not quite reach the arms. The posterior face of the thighs is covered with granular scales, excepting a large patch near the groin rising to near the upper surface of the thigh, which are imbricated and rhomboidal, similar to those of the inferior surface of the thigh. The scales of the upper part of the toes are smooth, those beneath have one or two slight carinæ.

This species varies very much in colour; three or four tolerably distinct varieties may be noticed, but they often pass more or less into each other. As I have only one of these varieties in my possession, I quote the following description from Mons. Bibron's account of the specimens in the French Museum.

Var. A. Les parties supérieures sont grises, ou bien d'un brun marron plus ou moins clair. Il règne au long du cou et du dos quatre series de taches noires, bordées de blanc en arrière. La queue et les membres offrent des bandes transversales anguleuse, d'un teint marron noirâtre, alternant avec des bandes semblables mais de couleur blanche. Les régions inférieures aussi sont blanches, excepté la gorge, qui est parcourue par des raies confluentes brunes. D'autres raies d'un brun marron sont imprimées verticalement, sur les lèvres.

Var. B. Cette variété se distingue de la precedente, en ce que le dessus de ses membres est ponctué de noirâtre, et que les quatre séries de taches qui ornent le dos de la première variété sont appliquées ici sur un fond fauve jaunâtre. Puis la gorge est verdâtre et le ventre noir, marbré de blanc.

Var. C. Le dessus du corps est uniformement peint d'un vert olive. Le dessus du cou, le milieu de la poitrine et celui du ventre sont d'un noir profond.

DIMENSIONS.

	Inches.	Lines.
Length of the head...................	0	7
of the body	2	8
of the tail	4	2
Total length....	7	7
Length of anterior extremity	1	2
of posterior extremity	2	0

This species agrees with *Pr. Darwinii* and *Weigmannii,* and in some degree with *Pr. Kingii,* in having a portion of the posterior face of the thighs covered with imbricated scales. This is a character, which although existing in all those which I have named, is found to obtain in very different degrees; in *Tr. Weigmannii* being very distinctly marked, and in *Tr. Kingii* very slightly so, and in some specimens scarcely notable. It is probable, that var. B., and possibly A. also, of Mons. Bibron, may be *Pr. Kingii;* but I have not had the opportunity of ascertaining this from the actual examination of the specimens.

Found by Mr. Darwin at Port Desire, and at Santa Cruz, in Patagonia.

PROCTOTRETUS CYANOGASTER.

PLATE V.—FIG. 2.

Squamis capitis neque imbricatis nec carinatis ; temporum imbricatis, subcarinatis, margine rotundato ; aurium margine anteriore simplici ; squamis dorsalibus rhombeis, laxis, carinâ posticè acutâ ; femorum facie posteriore omninò granulosâ ; corpore suprà olivaceo, fascia utrinque longitudinali flavescenti ; abdomine cœruleo.

Proctotretus cyanogaster, Bibr. l. c. p. 273.

Habitat, Valparaiso, and Valdivia, Chile.

DESCRIPTION.—It has been well observed by M. Bibron that this species offers at first sight somewhat the general aspect of the genus Algira ; the acute points of the dorsal and lateral scales and the general form giving very much that appearance.

The head is of moderate size, somewhat deflexed ; the scales moderate, flat and smooth ; those of the temples are slightly carinated, imbricated and rounded: those of the sides of the neck small, not granular, but rhomboidal and imbricated. There is but a single series of oblong scales between the labial ones and the orbit. The margin of the ear is entire and simple ; the scales of the back are lozenge-shaped, the carina of moderate height but prolonged into an acute point. Amongst those of the sides of the neck and belly are a few which are notched at the margin.* The scales of the posterior surface of the thigh are wholly granular.

The proportions of the limbs vary in the two sexes. In the male they are considerably longer than in the female. In the latter the posterior extremity when placed against the side extends only to the arm, in the former it reaches to the ear. The ground colour of the back is chesnut brown or greenish brown, with a bright metallic green glance in certain lights; there are two light buff longitudinal fasciæ running the whole length of the body; the under parts are of a bright metallic blue colour. Mr. Darwin states that in one specimen there were emerald spots on the sides, which did not exist in another individual. This may possibly be a sexual peculiarity.

DIMENSIONS OF A MALE SPECIMEN.

	Inches.	Lines.
Length of the head..............	0	6
of the body	1	9
of the tail	3	3
Total length....	5	8
Length of anterior extremity	1	0
of posterior extremity	1	5

Found by Mr. Darwin at Valparaiso, and at Valdivia; the former is a very dry rocky country, with a scanty vegetation; whereas the latter is nearly level, covered with the thickest forest, and the climate exceedingly humid.

PROCTOTRETUS KINGII, N.S.

PLATE VI.—FIG. 1, 2.

Squamis capitis neque imbricatis nec carinatis ; supralabialibus in serie unicâ ; aurium margine anteriore granuloso; interdum unidentata ; squamis, dorsalibus carinatis, posticè acuminatis, femorum facie posteriore præcipuè granulosâ, sed portione parvâ, caudam versus, squamis parvis, rotundatis imbicratis tectâ.

Habitat, Port Desire in Patagonia.

DESCRIPTION.—General form robust and full; the head short, thick, and passing into the neck without any distinct contraction; the muzzle rounded. Scales of the head larger in proportion than in *Pr. Fitzingerii*. Ear large, oval, with the anterior margin granular, sometimes slightly toothed. Scales of the temples of moderate size, imbricated, smooth, somewhat raised. A single range of oval moderate-sized scales between the labial and the orbital scales on the

* This is contrary to the character given by M. Bibron, who states that the whole of these are entire.

sides of the neck, above and behind the shoulder small, granular, and some of them having a
minute pore. Scales of the back of moderate size, larger than in *Fitzingerii*, rhomboid, having
a rather prominent carina, and terminating in a distinct point. Scales of the under parts
smooth and rhomboidal. The posterior surface of the thighs is for the most part granular, but
a small portion near the groin is covered with larger imbricated scales; to a much smaller
extent, however, than in *Fitzingerii*, and other species which possess this character.

The general colour of the upper part of this species is a rich dark brown, with whitish
transverse bands and spots, having a black margin. I have figured in fig. 2, of Plate VI., a
remarkable variety in which the bands are alternate black and white, and a broader and a
narrower longitudinal fascia of a yellowish-white colour, run the whole length of the body on
each side. The under parts are yellowish-white, with dark or almost black spots; under the
throat bluish-gray with white spots.

DIMENSIONS.

	Inches.	Lines.
Length of the head................	0	8
of the body	2	3
of the tail	3	4
Total length....	6	5
Length of anterior extremity	1	1
of posterior extremity	1	4

This species much resembles *Pr. Fitzingerii* in many of its characters, as
well as in its size. It may, however, be at once distinguished from it by the character of the scales of the back, which in this species are very distinctly carinated,
of a rather elongated form, and pointed at the extremity; whereas in the other
they are shorter, smaller, the carina is very slight, almost indistinct, and the posterior extremity is obtuse.

The tail in the larger figure of our plate is deformed, having been renewed.
The specimen figured at (2,) in the same plate, is so remarkably distinct in the
colours and marking, as to lead me to suppose that it may possibly be a different
species.

PROCTOTRETUS DARWINII.

PLATE VII.—FIG. 1, 2.

Corpore subdepresso; capite squamis numerosis, parvis, subelevatis, lævibus non imbri-
catis; aurium margine anteriore integro; temporibus colloque granulatis; serie
unicâ squamarum supralabialium; facie posteriore femorum partìm granulosâ,
partìm squamis imbricatis tectâ.

Habitat, Bahia Blanca, Northern Patagonia. Mr. Darwin.

DESCRIPTION.—The general form of this new species is similar to that of *Pr. Weigmannii*, but less elongate and somewhat more depressed, resembling in general appearance some of the forms of the genus Sceloporus. Head covered with rather small and consequently numerous scales, slightly elevated, and separated from each other by distinct and deep lines. A single series of small scales between the labial scales and the orbit. The exterior margin of the ear is entire and even. Scales of the temple and at the sides of the neck wholly granular, the latter very small. The scales of the back are small, flat, with a very low carina, and not pointed at the posterior extremity. There are about twenty rows of dorsal scales. The posterior surface of the thighs is granulated, excepting a small patch near the tail of imbricated scales, similar to those of the inferior surface, as in *Pr. Weigmannii*. The tail is of moderate length, and the scales which cover it are short, depressed, and obtuse in comparison with those of several other species. The pre-anal pores, which are peculiar to the male, are about ten in number.

The general colour is gray, with two light longitudinal lines on each side, and a row of black spots along the inner margin of the dorsal ones. The under surface is nearly white, with black dots under the throat.

The anterior legs, when stretched backwards against the side, reach about two-thirds towards the thigh; and the posterior when stretched forwards, extend to the shoulder.

DIMENSIONS.

	Inches.	Lines.
Length of the head	0	5
of the body	1	6
of the tail	3	4
Total length....	5	5
Length of anterior extremities	0	8
of posterior extremities	1	2

It is at first sight extremely difficult to distinguish this species from younger individuals of *Pr. Fitzingerii*, from which, however, it differs in the more linear form of the supralabial scales, in the absence of imbricated scales on the lateral fold of the neck, the more entire margin of the ear. In the existence of a patch of larger imbricated scales on the posterior surface of the thighs, it resembles *Pr. Weigmannii*; from which, however, it may be at once distinguished by the single row of supralabial scales, the later species having a double row.

PROCTOTRETUS WEIGMANNII.

PLATE VIII.—FIG. 1, 2.

Capite squamis lævibus non imbricatis tecto; auribus rotundis margine anteriore minutè granulato. Seriebus duabus squamarum supralabialium. Femorum facie posteriore partìm granulosâ, partìm squamis minutis imbricatis tectâ.

Proctotretus Weigmannii, Bibr. l. c. p. 284.

Habitat, Northern Patagonia and La Plata.

DESCRIPTION.—Head rather short, covered with numerous slightly raised scales, not carinated nor imbricated ; snout rather obtuse and slightly rounded, nostrils semicircular. Two series of very small scales between the labial scales and the orbit. Ear of moderate size, the anterior margin furnished merely with minute granular scales. Scales of the temples flat and smooth. Fold on the sides of the neck distinctly marked, anteriorly bifurcated; the remainder somewhat waved. Sides of the neck granulated scales of the whole of the upper and back parts of the body and tail of moderate size, the carina little elevated and the point but slightly prominent. The scales of the inferior parts of the neck and body are smooth, polished, and imbricated, those towards the sides of the neck minutely emarginated. The scales of the limbs resemble those of the body; those of the upper surface being carinated and those beneath smooth. The posterior face of the thighs is generally covered with granular scales, but there is on this part near the tail, a distinct patch of imbricated scales resembling those of the inferior surface of the thighs, a character by which this species may at once be distinguished from all others. The anterior extremity placed against the side reaches about two-thirds of the distance towards the groin; the posterior extremity reaches forward to the shoulder.

COLOUR.—The back and sides are brownish gray, with a yellow longitudinal band on each side of the back, separating transverse black or dark brown bands of various size and form; and there is in most on each side a smaller interrupted yellow band. The under parts generally of a yellowish white, in some individuals sparsely dotted with black. Mr. Darwin says of some individuals of this species that they have " an orange-coloured gorge, and faint stripes of blue," also " ash-grey with dark brown marks and specks of orange and blue."

DIMENSIONS.

	Inches.	Lines.
Length of the head	0	5
of the body	1	5
of the tail	2	2
Total length....	4	2
Length of anterior extremity	1	1
of posterior extremity	0	8

This species was found by Mr. Darwin at Bahia Blanca and at Rio Negro, on the northern confines of Patagonia, and at Maldonado, near the mouth of the Rio Plata.

PROCTOTRETUS MULTIMACULATUS.

PLATE IX.—FIG. 1.

Corpore subdepresso ; capite squamis numerosis parvis tecto ; auribus parvis, margine lævi ; seriebus quatuor squamarum supralabialium ; squamis temporum imbricatis ; collo granuloso ; femorum facie posteriore partìm granulosâ, partìm squamis imbricatis tectâ.

Proctotretus multimaculatus, Bibr. l. c. p. 291.

Habitat, Bahia Blanca, Northern Patagonia.

DESCRIPTION.—The body depressed and wide—the head triangular, the muzzle rather acute. Nostrils prominent and nearly round. Scales of the head very small and numerous ; those of the temple rhomboidal and imbricated. There are four series of small irregular scales between the labial and sub-orbitar. The sides of the neck are wholly granular ; the scales of the body very small ; those of the upper parts rhomboidal, flat with very low carina, and obtuse at the apex—beneath they are also small and rhomboidal ; the posterior surface of the thighs is granular, but, as in some other speeies, there is, near the groin, a distinct patch of imbricated scales like those of the inferior surface. The tail is broad to some distance from the origin, and then tapers to the extremity. The scales of the tail are rather small, short and obtuse.

The anterior extremity placed against the sides reaches about two-thirds the distance towards the posterior, and the latter reaches forwards to the shoulder.

The ground colour of this speeies is gray, with numerous small black spots, some of which are bordered with white. The under parts are white, and in one specimen in Mr. Darwin's collection there are on the belly numerous distinct small blaek spots. His description of the colours is as follows:—" Colours above singularly mottled. The small scales are coloured brown, white, yellowish red and blue, all dirty, and the brown forming symmetrical clouds. Beneath white, with regular spots of brown on the belly."

DIMENSIONS.

	Inches.	Lines.
Length of the head	0	8
of the body	1	8
of the tail	3	0
Total length ...	5	6
Length of anterior extremity	1	0
of posterior extremity	1	6

Found at Bahia Blanca, on the northern confines of Patagonia. The following remarks of Mr. Darwin on the habits of this species are very interesting. " In its depressed form and general appearance it partakes of some of the characters

D

of the Geckos. Its habits are singular. It lives on the dry sand of the beach, at some distance from the vegetation, and the colour of the body much resembles that of the sand. When frightened it depresses its body, stretches out its legs, and closing its eyes tries to escape detection. If pursued it buries itself with great quickness in the sand; but as its legs are short, it cannot run very swiftly."

Proctotretus pectinatus.

Plate IX.—Fig. 2.

Capite squamis subæqualibus, rhomboideis, imbricatis, carinatis tecto.

Proctotretus pectinatus, Bibr. Hist. Rept. IV. p. 292.

Habitat, Patagonia.

Description.—The scales of the head are narrow, closely imbricated, strongly but not acutely carinated, and the anterior ones arranged in somewhat of a radiating direction from the muzzle. There is but a single series of scales between those of the upper lip and the orbit, and these, together with all the scales about the head, partake of the carinated and elongated character already described. A single strong triangular scale and two smaller ones are placed on the anterior margin of the ear, which is narrow, oval and reniform. The scales of the temples and sides of the neck are rhomboidal, acute, carinated and imbricated. There is a longitudinal fold on each side of the neck and a transverse one anterior to the shoulder, behind which is a deep depression. The scales of the back and side are prominently and acutely carinated, those of the central line being rather more prominent than the others; and above this there is on each side a marked longitudinal lateral crest extending from beneath the eyes to the base of the tail. The scales constituting these crests are very prominent, narrow and acutely carinated. The scales of the belly are also imbricated and rhomboidal, but flat; those of the under surface of the hands and feet are carinated; and those of the toes have three carinæ. The body is somewhat depressed as is the tail at its commencement, becoming more rounded and rather abruptly smaller at some distance from its origin. The fore-foot reaches to about two-thirds of the distance from the shoulder to the side, and the hinder extremity thus placed extends to the shoulder.

The colours of this most elegant of all the species of the genus are very beautiful. " This is the most beautiful lizard," says Mr. Darwin, " I have ever seen ; the back has three rows of regular oblong marks of a rich brown, the other scales symmetrically coloured either ash or light brown; many of them of a bright emerald green; beneath pearly, with semilunar spots of brilliant orange on the throat." I find in the specimens I have examined that the pectinated lateral crests are white, and the brown oblong marks of the back are bordered with a similar colour. There are always three white transverse lines across the head.

DIMENSIONS.

	Inches.	Lines.
Length of the head	0	7
of the body	1	7
of the tail	3	1
Total length	5	5
Length of anterior extremity	1	0
of posterior extremity	1	5

This species, as has been observed by M. Bibron, who first described it, may be at once distinguished from every other by the character of the scales of the head, which, instead of lying flat, with the edges in contact, are all of them imbricated and carinated. Another obvious distinguishing character, is the narrow line of prominent scales running the whole length of the body on each side, forming a sort of *pectinated* lateral crest, from which circumstance it has derived its name.

Found by Mr. Darwin, at Bahia Blanca, and Port Desire in Patagonia.

Genus—DIPLOLÆMUS. *Bell.*

Caput *breve, latum, subtriangulare.* Aures *parvæ, ovatæ, margine lævi.* Nares *magnæ, rotundæ.* Collum *infrà transversè, ad latera longitudinalitèr plicatum.* Corpus *subdepressum, non cristatum.* Cauda *teres, breviuscula, lævis.* Pedes *breves, robusti.* Squamæ capitis *numerosæ, parvæ, rotundatæ, non imbricatæ*— corporis *atque* caudæ *suprà minimæ, læves, convexæ, paulò imbricatæ, infra læves, planæ.* Pori femorales *et* præ-anales *in utroque sexu nulli.* Dentes palatini *nulli.*

The new genus which I have thus defined, resembles very closely, in most of its characters, the genus *Leiosaurus* of Bibron ; from which, indeed, it scarcely differs, excepting in the absence of palatine teeth, and in the form of the suborbitar plates, which in *Leiosaurus* are all distinct, and of nearly equal size : whereas, in the present genus, three of these are united to form one plate, resembling that in *Proctotretus*, and some other *Agamidæ.* In other respects the genera are very closely allied ; but the existence or non-existence of palatine teeth, is a character of so much importance, that it appeared to me,—and in this opinion I am supported by M. Bibron, who examined the specimens with me,—that they should be considered as distinct. Both the genera are natives of South America. Of *Leiosaurus Bellii* (Bibr.) the only known specimens were presented to me by Capt. King, who obtained them during his survey, from whom also I obtained specimens of one of the species of the present genus, *D. Bibronii.*

DIPLOLÆMUS DARWINII. *Mihi.*

PLATE X.

Squamis capitis convexis; caudâ, corpore cum capite longiore.

Habitat, Port Desire, Patagonia.

DESCRIPTION.—Head short, almost equilaterally triangular, rising obliquely from the muzzle to the vertex, then flattened. Nostrils large, round, each placed in front of the supra-orbital crest, and in a line between it and the centre of the muzzle. The ears are small, oval, the margin simple, and the membrana tympani superficial. The neck is considerably contracted; it has a longitudinal fold on each side formed by the confluence of two others, one of which arises from behind the angle of the mouth, and the other from above the ear, which is, as it were, enclosed between them; they coalesce a little behind the ear. There is also a distinct transverse fold on the throat, very similar to that in *Leiosaurus Bellii.* The body is moderately thick, somewhat depressed, and without the slightest appearance of a longitudinal crest, or any elevation along the median line. The tail is somewhat longer than the head and body, nearly round and tapering almost evenly from its origin to the apex. The fore legs are short and moderately robust, the toes short, nearly equal; the hinder legs moderately long. The former when placed against the sides, do not reach the thighs by nearly a third of the distance between the two limbs; the latter when directed forwards, just reach the axillæ. The cloacal covering is semilunar, turgid, and the margin quite simple.

Scales covering the upper surface of the head numerous, rounded, and considerably elevated; those between the two supra-orbital semicircles are in a double series. The occipital plate is oval, raised from the margin, hollowed immediately around the centre which is again raised like a minute tubercle. Above the labial scales, is a series of equal, rounded, oblong scales, and between these and the principal suborbital is a single series of smaller ones. Scales of the whole of the upper and lateral parts of the neck and body extremely small, slightly elevated, passing at the sides into a flatter and more expanded form. Those of the whole of the under parts are quite flat and imbricated. Beneath the anterior parts of the lower jaw, and behind the broad mental scales, are a series of flat, hexagonal scales on each side, passing backwards and outwards, the front pair large and oblong and the others diminishing by degrees. The scales of the throat are very small, those on the fold larger and acutely rhomboidal. The scales of the anterior part of the belly are also rhomboidal and those of the posterior portion hexagonal or nearly quadrate. The tail is covered by scales disposed in whorls, those on the median line beneath being larger than the others. Beneath each toe is a series of transverse hexagonal imbricated scales.

The colours and markings of this species are very difficult to be described, on account of the great irregularity of their disposition. The ground colour of the head is yellow, passing into grey on the back part. The anterior part has several small spots of a dark brown colour, and there is a larger one on each orbit, another between the eye and the ear, and others on the back part of the head extending to the neck. The middle of the back is reddish yellow, on each side bluish gray, passing beneath into yellowish white. A series of very irregular transverse spots cross the yellow median portion of the back, and there are others on the sides; and these two series becoming confluent on the tail, form, with the yellow ground, alternate half rings of the two colours. The upper part of the legs has similar bands. The whole of the throat, belly, and inferior surface of the limbs and tail are yellowish white. There are numerous small blackish spots over these parts which are more distinct and linear on the throat, and becoming paler, smaller and round on the belly.

DIMENSIONS.

	Inches.	Lines.
Length of the head	1	0
of the body	2	2
of the tail........................	3	8
Total length......	7	0
Length of anterior extremity	1	1
of posterior extremity	1	6

Taken at Port Desire, on the coast of Patagonia.

DIPLOLÆMUS BIBRONII. *Mihi.*

PLATE XI.

Squamis capitis planis; caudá corpore cum capite breviore.

Habitat, Port Desire.

DESCRIPTION.—Head thick and clumsy, longer than it is broad, muzzle obtuse, supra-orbital arches slightly elevated. Nostrils as in the former species, in size, form, and situation. Ears sub-triangular, the margin simple. Neck considerably contracted, with a longitudinal fold on each side, and a distinct transverse fold on the throat. Body rather broad, slightly depressed, perfectly even, without any central crest or elevation. The tail is shorter than the head and body, slightly triangular at its base, tapering regularly to its extremity. Limbs of moderate length; the toes of each foot longer than in *D. Bibronii*, and those of the fore-feet more unequal, the third being the longest, then the fourth, the second, the fifth, and the first. The fore-legs placed against the side reaches to about two-thirds of the distance between the shoulder and thigh; the hinder foot placed in the same manner reaches to the axilla.

The scales of the head are quite flat, a character in which this species differs remarkably from the former, although in their number and arrangement they are very similar. The occipital scale is flat and hexagonal. Between the labial scales and the suborbital, there are, in addition to the regular series of larger supralabial scales, at least three distinct series of smaller ones; whereas in *D. Darwinii* there is but one.

The scales of the temples, the neck, the body, the limbs and the tail, are similar to those of the former species in general form and arrangement, excepting that they are smaller and less elevated. Those beneath the anterior part of the lower jaw are much smaller; but the rest on the under parts are similar to the former.

The head is of a dull light-brown colour, with a few obscure darker spots. The general ground colour of the back is " bluish gray, tinged with rust colour;" there are five transverse bands across the back, which are composed principally of numerous, close, small, dark-brown spots, on a bluish-gray ground, darker than the intervals, and without any red tinge; and each band is marked on the posterior margin with strongly defined semilunar indentations, bordered with yellowish-white, or bright yellow. These bands are continued on the tail, where they become half-rings.

DIMENSIONS.

	Inches.	Lines.
Length of the head	1	2
of the body	2	9
of the tail	3	5
Total length ...	7	6
Length of anterior extremity	1	4
of posterior extremity	2	1

Genus—AMBLYRYNCHUS. *Bell.*

AMBLYRYNCHUS DEMARLII. *Bibr.*

PLATE XII.

Cristâ supra cervicem elevatiore, supra dorsum humiliore; tuberculis verticalibus sub-depressis, occipitalibus conicis; caudâ tereti.

Amblyrynchus Demarlii. Bibr. Hist. Rept. IV. p. 197.

This species was first described by Mons. Bibron in the "Histoire des Reptiles," and so fully as not to require any detailed account of its characters here. It has not, however, hitherto been figured, and it is thought very desirable to embrace so good an opportunity of giving a representation of so interesting an animal. Its most important structural peculiarities will be alluded to in the account of the next species, which is an aquatic form, whilst the present is strictly terrestrial. The toes are long, compared with those of the other, and so unequal as to constitute essentially an ambulatory form.

By Mr. Darwin's observations we are now enabled fully to confirm Mons. Bibron's suggestion, that this species was from the Galapagos, and to establish the genus as strictly appertaining to that curious and interesting locality.

AMBLYRYNCHUS CRISTATUS. *Bell.*

Cristâ supra humeros humiliore.; digitis ferè equalibus subpalmatis ; caudâ compressâ.

Amblyrynchus Cristatus. Bell, Zool. Journ. 1825, p. 195.   Tab. Supp. XII.   Bibr. Hist. Rept. IV. p. 204.

I established the genus *Amblyrynchus* nearly eighteen years ago, from a stuffed specimen of the present species, which had been obtained by Mr. Bullock, Jun., in Mexico. I had never seen another specimen, until Mr. Darwin brought home a young one from the Galapagos, in excellent preservation in spirits, and thus established its true habitat, and enabled me to correct those errors in my description which arose from drying and bad stuffing. Mons. Bibron also took his description from my specimen, and thus necessarily fell into the same mistakes, of which the most important are those which relate to the form of the tail, and the structure of the feet. Thus the tail is described as " round, excepting towards the extremity, where it is flattened at the sides," whereas it is in fact much compressed throughout its whole length ; and with regard to the toes no mention is made of their being partially united by a web or fold of skin, which is the case both on the anterior and posterior feet. These two characters so obviously point out a power of swimming, that the aquatic habits of the species might at once have been predicated, and it is exceedingly interesting to find, from Mr. Darwin's observations, that such is really the case. We have, therefore, two distinct forms —distinct equally in their structure and in their habits—in the two species now described ; the one, *A. Demarlii,* being truly terrestrial, with lengthened, unequal, and distinctly separated toes and a round tail, and the present species as truly amphibious, having short, nearly equal and webbed toes, and a compressed tail.

A very interesting account of their habits, &c., is given by Mr. Darwin in his delightful Journal of the Voyage of the Beagle, p. 466 to 472, to which the reader is referred, and which exactly accords with the peculiarities of their respective structure just alluded to.

It is remarkable also, that whereas *Amblyrynchus cristatus* inhabits the coasts of all the islands, the other species is found only in the central portion of the group.

Genus—LEIOCEPHALUS. *Gray.*

Leiocephalus Grayii.

Plate XIII.—Fig. 1.

Cristâ dorsali elevatâ; caudâ sub-compressâ; squamis ventralibus rhomboideis, lævibus; margine anteriore meatus auditorii quadridentato; squamâ occipitali magnâ.

Habitat, Galapagos Archipelago.

Description.—Head, viewed from above, forming a nearly equilateral triangle, covered with irregular slightly raised scales. Supra-orbital ridge prominent, and covered with a series of elongated and imbricated scales. Occipital plate large, pentagonal, notched at its posterior margin. The anterior margin of the auditory passage is strongly quadridentate, from the existence of four long and rather narrow scales. Scales of the temple obtusely carinated, not imbricated; those of the back strongly and acutely carinated and disposed in numerous rows, converging backwards towards the dorsal crest. Ventral scales rhomboidal, not carinated. Dorsal crest elevated, composed of flat vertical scales, so closely placed as to constitute an almost continuous line, extending from the neck to the end of the tail. Tail somewhat compressed at the base, becoming nearly round towards the middle. Scales beneath the feet and toes carinated.

Colour.—The colour of this species is thus stated in Mr. Darwin's notes:—" Upper part clove brown, passing into black brown with black spots. Sides slightly tinted with orange; some of the scales of the crest near the head are white; belly nearly white; the whole of the throat before the fore legs glossy black. This is the most common variety in the Archipelago. The black spots are not unfrequently placed in waved transverse bars, and are sometimes arranged longitudinally.

DIMENSIONS.

	Inches.	Lines.
Length of the head	0	9
of the body	2	8
of the tail	5	8
Total length	9	5

Of this species, one of the most beautiful in the whole order of Saurians, Mr. Darwin obtained numerous specimens, one only of which is fully adult. In the younger individuals the dorsal crest is low and almost inconspicuous. It differs very materially from either of the two species previously described, and I have dedicated it to Mr. Gray, who first distinguished the genus. Mons. Bibron, unaware that Mr. Gray had already constituted the genus under the name *Leiocephalus*, named it *Holotropis*. I have, however, retained the former name, as having the claim of priority.

It constitutes one of the numerous interesting novelties obtained by Mr. Darwin in the Galapagos. The specimens, which are of various ages, were taken in Chatham Island and in Charles Island.

Genus—CENTRURA. *Bell.*

Caput *breve, triangulare.* Aures *magnæ, anticè cutis plicâ, haud dentatâ, partim celatæ.* Nares *magnæ, rotundæ.* Gula *transversè subplicata.* Collum *atque* corpus *haud cristata ; hoc depressum, latum, cute longitudinalitèr plicatâ.* Cauda *teres, basin versus subdepressa, squamis fortibus spinosis verticillatis.* Squamæ capitis *numerosæ, parvæ, rotundatæ, non imbricatæ*—corporis *minimæ, rotundæ, subconvexæ, læves.* Pori femorales *et* præ-anales *nulli.* Dentes palatini.

The propinquity of this genus both to *Oplura* and to *Doryphorus* is very obvious. It differs, however, from both in several structural characters. From the former in the absence of denticulations on the anterior margin of the ear, and of a nuchal crest; from the latter in the presence of palatine teeth. Its place is probably between these two genera.

CENTRURA FLAGELLIFER. *Mihi.*

PLATE XIII.—FIG. 2.

DESCRIPTION.—Head almost equilaterally triangular, the muzzle rounded; scales of the head small, nearly equal, rounded, not imbricated, those of the temples subconical; nostrils round, large, confined to the nasal scales. Ears rather large, the tympanum lying beneath the surface, and partly concealed by an anterior fold of skin, which is not denticulated, as in *Oplura.* Skin of the neck folded at the sides, that of the body flaccid, and with strongly marked lateral folds, extending from the shoulder to the thigh. Scales of the neck and back very small, round, slightly convex, very smooth. Skin of the throat rugose, with a transverse pectoral fold not very strongly marked. Scales of the throat similar to those of the back; those of the belly broader and less convex ; all perfectly smooth. Tail about the length of the head and body, flattened at the base, then round, surrounded with strong spinous verticillated scales, of which there are about fifty circles; beneath smooth. Legs of moderate length, strong, covered with small conical, imbricated scales. The toes compressed towards the extremity, and terminated with a strong, short, compressed nail.

COLOUR.—The colour can only be partially described, as the specimen has been long in spirits. It

E

is of a dark brown colour above, with darker, obscure markings on the body. About the head
are traces of green. The tail and limbs are rich brown, and the under parts dull, pale
fuscous.

DIMENSIONS.

	Inches.	Lines.
Length of the head	0	8
of the neck	0	4
of the body	2	5
of the tail	3	7
Total length	7	4
Length of anterior extremity	1	4
of posterior extremity	2	0

FAMILIA—GECKOTIDÆ. *Gray.*

GECKOTIENS OU ASCALABOTES. *Bibr.*

GENUS—GYMNODACTYLUS. *Spix.*

GYMNODACTYLUS GAUDICHAUDII. *Bibr.*

PLATE XIV.—FIG. 1.

*Squamá mentali impari pentagoná, scutiformi; squamarum labialium inferiorum
paribus quinque, superiorum paribus sex ; caudá medio crassiore.*

Gymnodactylus Gaudichaudii. Bibr. Rept. III. p. 413.

This species was first brought from Coquimbo by Gaudichaud, after whom it
was named by Mons. Bibron, who described it in his work ; but as one specimen
alone exists in the French National Collection, and as the species has never
been figured, it is thought desirable that a figure should be given in the
present work. The characters above given sufficiently distinguish it from all
other species ; but for a detailed description, the reader is referred to the " Histoire des Reptiles" above quoted.

DIMENSIONS.

	Inches.	Lines.
Length of the head	0	5
of the body	1	4
of the tail	2	3
Total length	4	2

The specimens brought home by Mr. Darwin were from Port Desire, in Patagonia, and the following observations occur in his MS. notes :—" Centre of the back yellowish brown, sometimes with a strong tinge of dark green ; sides clouded with blackish brown ; in very great numbers under stones ; makes a grating noise when taken hold of ; after death loses its darker colours.

" A specimen being kept for some days in a tin box, changed colour into an uniform grey, without the black cloudings. I thought I noticed some change after catching and bringing home these animals, but could observe no instantaneous change."

I have considered these specimens as belonging to the species to which I have assigned them, because they exactly agree with Mons. Bibron's description. It is, however, very possible that an opportunity of comparing them with those obtained by Gaudichaud, would show them to be distinct, as it rarely happens that the same species of reptile is found on the opposite sides of the American Continent.

GENUS—NAULTINUS. Gray.

NAULTINUS GRAYII.

PLATE XIV.—FIG. 2.

Omninò viridis ; fronte subconcavo ; squamulis capitis planis.

DESCRIPTION.—Head thick, swollen across the posterior part, concave between the eyes, and forwards nearly to the snout, which is rounded. Scales of the head larger towards the fore part, nearly flat. Eyes round, large ; ears longitudinally oval. Body covered with small nearly equal scales. Tail round, one-fifth longer than the body. Limbs short, the anterior, when placed against the side, reaching but little more than half way to the thigh ; the posterior reaching about two-thirds the distance towards the shoulder. Toes short ; on the anterior foot the first is the shortest, then the second, the fifth, and the fourth ; on the posterior increasing in the same series ; all compressed towards the extremity, and all furnished with small curved close claws.

The colour is a fine green.

It was taken at the Bay of Islands, New Zealand. It lives on trees, and is said to make a laughing noise.

This species greatly resembles Naultinus Elegans * of Mr. Gray, of which a beautiful specimen is in the British Museum. Upon a comparison of the two,

* See Fauna of New Zealand, p. 203. Zool. Misc. p. 72.

however, I find that they differ in the following particulars. In the present species the head is concave between the eyes, and forwards nearly to the snout; in the other, this part is quite plain; the scales of the head in this species are flat; in the other they are convex. The colour of this species is uniformly green, whereas *N. Elegans* has several markings of a yellow colour, each distinctly bordered with black.

Familia — LACERTIDÆ.

Genus—Ameiva. *Cuvier.*

Ameiva longicauda. *Mihi.*

Plate XV.—Fig. 1.

Squamis supra-humeralibus, rhomboideis, imbricatis; subfemoralibus transversim hexa-gonis; abdominalibus in seriebus decem longitudinalibus dispositis; caudá, corpore cum capite plus quam dupló longiore, squamis medio carinatis, et ad margine sub-carinatis.

Habitat, Bahia Blanca, Northern Patagonia.

Description.—Head very narrow, much elongated and pointed, the vertex flattened; nostrils rather large, open, round, directed laterally, and placed in the centre of the naso-rostral plate; super-ciliary plates three in number, the central one the largest; suprahumeral scales rhomboidal, imbricated, not broader than long, in four series; those of the arm transversely hexagonal; the anterior surface of the thigh, and the inferior of the leg, covered with large hexagonal, some-what imbricated, scales; caudal scales above quadrate, longer than broad, with a strong medial carina, and the lateral margins slightly raised; beneath smooth; tail very long. Anterior ex-tremity placed against the body, reaching rather more than half way to the thigh; posterior extremity extending forwards nearly to the ear.

Colour.—The upper surface of this beautiful species is dark brown or blackish, with nine distinct white or yellowish longitudinal fasciæ extending through the whole length of the neck and body; tail with four of these lines. Under parts white.

DIMENSIONS.

	Inches.	Lines.
Length of the head	0	6
of the neck	0	3
of the body	1	4
of the tail	5	2
Total length	7	5
Length of anterior extremity	0	7
of posterior extremity	1	5

Found at Bahia Blanca by Mr. Darwin. The specimens are probably all of them very young; hence the longitudinal lines can scarcely be considered as permanent, as most species of the genus are beautifully lineated in the young state. The length of the tail, with its carinated scales, the general elegance of the form, the gracile form of the head, and the neat and distinct arrangement of the colours, render this one of the most beautiful species of this elegant genus.

The description of the colours given above, being from specimens which have. been long in spirits, it is necessary to state that Mr. Darwin has the following notice respecting one of them—" On the sides two dark red streaks; tail red."

FAMILIA—ZONURIDÆ.

GENUS—GERRHOSAURUS. *Weigmann.*

GERRHOSAURUS SEPIFORMIS. *Bibr.*

PLATE XV.—FIG. 2.

Scincus sepiformis, Schneid. Hist. Amph. II. p. 191. Merr. Syst. Amph. p. 70. n. 1.
Gerrhosaurus sepiformis, Bibr. Hist. des Rept. V. p. 384.

Corpore cum caudâ longo, serpentiformi; pedibus parvis; squamarum submaxillarum pari secundo contiguis; squamis dorsalibus magnis, subrectangularibus, striatis, in seriebus tredecem, et ventralibus in seriebus octo dispositis.

Habitat, Cape of Good Hope.

After a careful examination of the data from which the different synonyms of this species, and of *Gerrhosaurus flavigularis,* Bibr., have been derived, I am inclined to agree with this author, that the present is the true *Scincus sepiformis* of Schneider, and of Merrem, and not *Scincus flavigularis* as supposed by Wagler, Weigmann, and Gray. It is very fully described by Bibron in the " Histoire Naturelle des Reptiles," but it has not hitherto been figured. There is no notice of it in Mr. Darwin's notes, further than its having been obtained at the Cape of Good Hope.

FAMILIA—SCINCIDÆ.

GENUS—CYCLODUS. *Wagler.*

CYCLODUS CASUARINÆ. *Bibr.*

PLATE XV.—FIG. 3.

Aurium margine anteriore simplici ; squamis corporis lævibus, in seriebus xxiv *dispositis.*

" *Kèneux de la casuarina,* Cocteau, Tab. Synopt." (*v.* Bibr. Hist. Nat. des Rept. V. p. 749.)
Cyclodus casuarinæ, Bibr. l. c.

As I have not the work of the lamented Dr. Cocteau by me, I quote the above reference from M. Bibron's work, in which this species is fully described. It differs from the other species of this curious genus in many minute characters of the scaling of the head, but the most tangible and obvious distinctive character consists in the number of series of scales, which does not exceed twenty-four, all around the body, whilst in the others, they amount to thirty-four or thirty-eight. It would appear that it is liable to some considerable diversity in colour and markings. That which M. Bibron describes, has " the head of a yellowish grey, the whole of the upper part of the body olive grey, and the inferior part whitish grey." The specimen in the collection of the Zoological Society has the whole upper part of a brownish grey, with twelve black lines extending from the neck along the back and tail, corresponding with the sutures of the longitudinal series of scales. The under surface of the tail is marked by about thirty transverse, interrupted, black bands. The following is the description given by Mr. Darwin from his specimen when taken,—" Scales on the centre of the back light greenish brown, edged on their sides with black; scales on the sides of the body above greyer and with less black, below reddish : belly yellow, with numerous narrow, irregular, waving, transverse lines of black, which are formed by the lower margin of some of the scales being black; head above grey, beneath whitish." Mr. Darwin adds, that the motion of the body, when crawling, resembles that of a snake. It is not very active. Coleoptera and larvæ were found in its stomach. " It is common in the open woods near Hobart Town in Van Diemen's Land."

Classis—AMPHIBIA.

Ordo—ANOURA.

Familia—RANIDÆ.

Genus Rana.

RANA DELALANDII. *Bibr.*

PLATE XVI.—FIG. 1.

Dentibus palatinis in serie transversá, medio interruptá, dispositis; membris posterior-
ibus corpore cum capite dupló longioribus; pedibus posticis gracillimis, semipalmatis.

Rana Delalandii, Bibr. Hist. Rept. VIII. p. 388.

DESCRIPTION.—Head elongate, depressed. Eyes large, not prominent. Tympanum nearly round.
Palatine teeth in two simple series, commencing at the inner side of the anterior margin of the
posterior nares, and extending towards each other in a transverse direction, leaving between
them a space of about half the length of each. Tongue not quite as long as it is broad. Body
somewhat depressed, and with the head forming an almost uninterrupted ellipse. Skin of the
back, with several small longitudinal folds. Anterior legs, when placed against the sides,
reaching to the thigh. Fingers very slender, and of nearly equal length. Posterior limbs
fully twice as long as the head and body. Toes extremely long and slender, and connected
by a membrane by about half their length.

COLOUR.—The general colour of the upper parts is a rich brown, with darker brown and white
markings. A white median fascia extends the whole length of the head and body; another
fascia of the same colour and of very irregular figure on each side, passes backwards and
downwards from above the shoulders, and loses itself in the pale colour of the abdomen.
There are several smaller white lines and spots, and others of a dark rich brown, particularly a
large mark of the latter colour behind the eye, including the tympanum. The thighs and
legs are elegantly banded with similar colours. The under side is whitish.

DIMENSIONS.

	In.	Lines.
Length of the head and body	1	8
of anterior extremities	1	1
of posterior ditto	3	7

This species was first discovered at the Cape of Good Hope by M. Delalande,
and named after him by Mons. Bibron. Mr. Darwin found it in the same locality.
It is now figured for the first time.

RANA MASCARIENSIS.

PLATE XVI.—Fig. 2.

Dentibus palatinis in fasciculis binis obliquis distantibus, ad marginem interiorem narium posteriorum attingentibus; tympano circulari, mediocri; digitis posticis usque ad phalanges penultimas connexis: plantis tuberculo unico; cute dorsi lævi, longitudinaliter plicatâ; suprà fusco-rufescens, fasciâ longitudinali pallidâ.

Rana Mascariensis, Bibr. Hist. Rept. VIII. p. 315.

Habitat, the Mauritius.

This pretty species of the typical genus of the family was described by Bibron, but has not hitherto been figured. It was found in Mauritius, on swamps near the sea, by Mr. Darwin, who remarks on the extraordinary height of its leaps. It has also been found in the Seychelles, Madagascar, and the Island of Bourbon.

GENUS—LIMNOCHARIS. *Bell.*

Lingua *ovalis, integra, margine posteriore libero.* Dentes palatini *utrinque in fasciculis duobus dispositis, quorum alter ad marginem anteriorem narium interiorum, alter pone nares interiores, prope arcum maxillarem* Nasus *terminalis, truncatus, ultra labium productus.* Tympanum *conspicuum, circulare.* Cutis *omninò lævis.* Digiti anteriores *liberi,* posteriores *ad basin tantùm palmati.*

The genus *Limnocharis* is remarkable for the existence of palatine teeth in a part of the mouth in which they have never been observed in any other amphibian. Not only is there a small group or line of these contiguous with the anterior margin of the posterior nares,—a situation in which they are found in some other genera of *Ranidæ,* but there is also a group of them placed at some distance behind the posterior margin of these openings, and close within the rise of the maxillary arch. This genus, of which one species only is at present known, will probably be most naturally placed between the true *Ranæ* and certain of the *Cystignathi.*

LIMNOCHARIS FUSCUS. *Mihi.*

PLATE XVI.—Fig. 3.

Habitat, Rio Janeiro.

DESCRIPTION.—Head semi-oval, depressed, as broad as it is long. The muzzle truncated, extending beyond the lips, which it overhangs. Tongue oval, entire, free at the posterior margin. Palatine teeth in two parcels on each side; one consisting of very few at the anterior and inner margin of the posterior nares, the other behind those openings, in the angle formed by the maxillary arch and the orbits. Posterior nares large and oval. Tympanum conspicuous, nearly circular. Skin every where perfectly smooth, without glands or pores. Anterior legs of moderate length and size. The fore-arm rather longer than the upper arm. Fingers of moderate length, wholly detached. Hinder legs little more than one-third longer than body. The toes separate, excepting a slight rudiment of a connecting membrane at their base, which extends, though very narrow, along their sides, the extremity very slightly notched.

COLOUR of the upper part rich dark brown. The thighs lighter, obscurely banded with dark brown. Under parts pale blueish grey. The throat dotted with brown.

DIMENSIONS.

	In.	Lin.
Length of the head and body	1	4
of the anterior extremities	0	7
of the posterior extremities	1	8

Found in brooks at Rio Janeiro by Mr. Darwin, who states that it is infested with acari; and I observe, in the specimen under examination, several marks in the skin, from whence these have been taken.

GENUS—CYSTIGNATHUS. *Wagler.*

CYSTIGNATHUS GEORGIANUS. *Bibr.*

PLATE XVI.—Fig. 4.

Dentibus palatinis perpaucis, in fasciculis binis approximatis, pone nares posteriores; linguâ integrâ, oblongâ; tympano celato; pedibus posterioribus non palmatis.

> *Crinia Georgiana,* Tschudi Class. Batrach.
> *Cystignathus Georgianus,* Bibr. Rept. VIII. p. 416.

This species, which formed the type of Tschudi's genus *Crinia,* was separated by him from *Cystignathus* on account of the form of the tongue, the non-

F

appearance of the tympanum, the paucity of palatine teeth, and the total absence of an interdigital membrane on the hinder feet. These characters being either merely comparative or unimportant, were not considered by Bibron as sufficient to warrant a generic separation, and I have followed him in retaining the species amongst the *Cystignathi*. It was first discovered by Messrs. Quoy and Gaimard at King George's Sound, in Australia, where it was also obtained by Mr. Darwin. It is a beautiful species; the back being of a rich brown colour, with a pale orange fascia extending along the sides from the eye to the thigh, becoming bright orange on the flanks. Thighs and legs banded with rich deep brown and bright orange.

Genus—BORBOROCŒTES. *Bell.*

Lingua ovata, posticè libera, rotundata; anticè subacuminata. Dentes palatini *in fasciculis binis plùs minùsve obliquis, pone nares posteriores positi.* Tympanum *celatum.* Digiti anteriores *haud palmati ; posteriores ad basin tantùm cute connexi.* Glandulæ cutaneæ *nullæ.* Sacculi vocales *(maris) utrinquè sub lingua nascentes.*

The two species on which I have founded this genus approach so nearly to some species of *Cystignathus*, that it is not without hesitation that I determine on considering them as typical of a new generic form. The principal characters on which I have founded the distinction are the position of the palatine teeth, the form of the tongue, the concealment of the tympanum, the absence of glands and pores on the skin, and the connexion of the base of the hinder toes by a rudimentary palmar membrane. It is true that some of the species of *Cystignathus*, as that genus is at present constituted, agree with the present form in some or other of these particulars; but upon the whole they are sufficiently distinct; and in fact the genus *Cystignathus*, as left by M. Bibron, appears to me to stand in need of revision and dismemberment. The species constituting the genus now proposed, are however both new. The genus *Borborocœtes* will probably stand, in its natural affinities, between *Cystignathus* and *Cycloramphus*, from the latter of which it differs in the situation of the palatine teeth, in the degree to which the hinder feet are webbed, and the comparative length of the hinder legs. The two latter characters are of importance as indicating a difference of habit; and we find that *Cycloramphus* has proportionally short hinder limbs, with the toes extensively palmate, whilst in *Borborocœtes* the hinder legs are much longer, and the toes scarcely at all webbed. The former structure indicates a greater power of swimming, and the latter of leaping.

BORBOROCŒTES BIBRONII. *Mihi.*

PLATE XVII. FIG. 1.

Dentibus palatinis in fasciculis distantibus obliquis pone nares posteriores positis ; palmis bituberculatis.

Habitat, Chiloe and Valdivia.

DESCRIPTION.—Head depressed, the vertex slightly concave between the orbits ; front (space in-cluded between two lines drawn from the anterior corner of the orbits to the point of the nose) triangular and distinct. Nostrils lateral. Eyes rather prominent. Tongue broad, ovate, acuminated in front, behind entire and rounded, the posterior half and the sides de-tached. Palatine teeth in two oval parcels, direct obliquely backwards and inwards, and situated at some distance behind the line of the posterior margin of the nares. Tympanum concealed. Body rather depressed and short. Skin smooth and without pores or glands, excepting on the posterior and inferior surface of the thighs, where there are some small granular elevations. Fore legs two-thirds the length of the head and body. The fore-arm rather larger than the upper arm. The fingers entirely separated, the third considerably the longest. A small tubercle under each joint, and two on the palm near the wrist. Length of the hinder legs to that of the head and body as 5 to 3, or rather more. Toes connected only at their base. A small tubercle under each joint, and a very depressed one at the base of the inner toe.

COLOUR of the upper parts fuscous, with a lateral fascia extending from the orbit nearly to the thigh, of a dark-brown colour, bordered with whitish ; and another of an elongated triangular form on each ilium. Legs with transverse incomplete faciæ of the same colour. Under parts grey, with numerous brown dots.

DIMENSIONS.

	In.	Lines.
Length of the body and head	1	5
of the anterior extremities	1	0
of the posterior extremities	2	6

Taken at Valdivia and at Chiloe, in a thick forest, by Mr. Darwin.

Borborocœtes Grayii. *Mihi.*

PLATE XVII.—Fig. 2.

Dentibus palatinis in fasciculis subcontiguis paulò obliquis, pone nares posteriores positis; palmis non tuberculatis.

Habitat, Valdivia.

This species considerably resembles the former in most of its characters. The palatine teeth, however, form at once a certain and tangible distinction, and there are some minor points in which they differ, sufficient at a glance to determine them. The head in the present animal is broader than it is long; in the former the breadth is only equal to its length. The palms are in this species without conspicuous tubercles; in the other there are two, although very small.

In colour it differs much from the former. The general colour is a rich fuscous brown, rather paler beneath; the flanks, the throat and belly, and the whole of the thighs and legs, with various white markings, those of the throat and belly being the smallest. This species was found in the forest, in Valdivia.

Genus—PLEURODEMA. *Tschudi.*

I have thought it right to follow Tschudi in separating from the genus *Cystignathus* of Wagler, such species as have large and conspicuous lumbar glands, particularly as they all agree in possessing a much more bufonine aspect than the others. The discovery of three new species, all agreeing in these characters with *Pleurodema Bibronii* of Tschudi, increases the importance of the grounds upon which this separation is made.

Pleurodema Darwinii. *Mihi.*

PLATE XVII.—Fig. 3.

Dentibus palatinis paucis, minimis; linguâ subcordatâ, vix emarginatâ; glandulis lumborum magnis, rotundis, convexis; digitis posticis ad basin tantùm membranâ connexis; dorso sparsìm tuberculato-glanduloso; suprà pallidè virescens, maculis fusco-olivaceis.

Habitat, Maldonado.

DESCRIPTION.—Head triangular, rather broader than long. Muzzle rounded. Eyes slightly prominent. Tongue somewhat heart-shaped, scarcely emarginated behind. Palatine teeth very
few, and with difficulty perceptible, placed in two small groups between the posterior nares.
Body thick and broad, with numerous glandular tubercles scattered over the surface, principally on the anterior parts, and assuming somewhat of a longitudinal arrangement. Lumbar
glands large, round, and prominent. Legs robust and short. Toes of the fore feet wholly
separate, with a small tubercle under each joint, and two larger ones at the hinder part of the
palm. Hinder toes, with a rudimentary membrane at the base, a small tubercle under each
joint; the first and second toes very short. A conical tubercle at the inner, and another at
the outer side of the metatarsus.

COLOUR.—The upper surface is beautifully marbled with dark olive or black, on a light-green
ground; some of the markings assume somewhat of an ocellated form, and approach to a
symmetrical arrangement. The lumbar glands are more strongly coloured than the other
parts, the centre being black, and nearly surrounded by a bright line of very light green, or
nearly white. The thighs are numerously banded with the prevailing colours, and a tinge of
orange or red. Beneath pale; in some specimens blackish under the chin.

The aspect of this species is remarkably bufonine; and this character is in-
creased by the numerous glandular tubercles on the surface of the body, and
pores about the parotid region. It is, doubtless, similar in its habits to many of
the toads.

DIMENSIONS.

	In.	Lin.
Length of the head and body..................	1	4
of the anterior extremities	0	8
of the posterior extremities	1	7

It was repeatedly found by Mr. Darwin at Maldonado, near the mouth of
the river La Plata.

PLEURODEMA ELEGANS. *Mihi.*

PLATE XVII.—FIG. 4.

*Dentibus palatinis prominentibus, in fasciculis binis ovatis obliquis dispositis;
lingua rotundâ integrâ; glandulis lumborum ovalibus, valdè convexis; digitis
posticis haud palmatis; dorso tuberculato-glanduloso, fusco, nigro obscurè ma-
culato, fasciâ longitudinali pallidâ.*

Habitat, Valparaiso, Valdivia, and Archipelago of Chiloe.

DESCRIPTION.—Head semi-elliptic, as broad as long. Muzzle rounded. Eyes very slightly prominent. Tongue large, round, entire, very thick. Palatine teeth prominent, disposed in two

oval groups, extending obliquely backwards and inwards, but separated by a considerable interval. Body somewhat depressed and elongated, with many prominent glandular tubercles, and with pores about the parotid region. Lumbar glands of moderate size, of an elongated oval form, and very convex. Legs rather slender, the anterior feet with the third toe considerably the longest; a small tubercle under each joint of all the toes, and several small inconspicuous ones on the palm; hinder legs rather elongated, the toes long, particularly the fourth, the first very short; a small tubercle under each joint; the inner metatarsal tubercle prominent, the outer one inconsiderable.

COLOUR.—The markings of this species are very elegant and striking. The ground colour of the upper parts is a rich brown, with darker cloudings and marks; a light yellowish longitudinal line running all the length from the nose to the extremity of the body, a very irregular fascia on each side of the same colour enclosing a brown oblong spot on the upper lip, another just behind the tympanum, and two others on the sides; there is also a brown fascia from the extremity of the nose to each eye; the lumbar glands are black and yellow, distinctly marked. The limbs are obscurely banded with brown and pale yellowish. The colours in some specimens are more obscure than in that figured, and they appear to lose their clearness with age.

The following are the colours of the brighter individuals according to the observations of Mr. Darwin :—" Yellowish and broccoli-brown, with darker brown marks; broad medial dorsal line, pale gallstone yellow; lumbar glands saffron yellow and jet black." Another specimen was " ash-grey with blackish brown marks."

DIMENSIONS.

	In.	Lines.
Length of the head and body	1	8
of the anterior extremities	1	0
of the posterior extremities	2	4

The general habit of this species is much more in accordance with its relation to the *Ranidæ* than that of the other species of the genus. Its general form is more elongated and depressed, and the limbs, particularly the hinder ones, are longer in proportion to the body. It is certainly very near *Pl. Bibronii* of Tschudi, but still undoubtedly distinct.

PLEURODEMA BUFONINUM. *Mihi.*

PLATE XVII.—Fig. 5.

Dentibus palatinis prominentibus, in fasciculis binis ovalibus, obliquis, dispositis; linguâ subcordiformi, subemarginata; glandulis lumborum maximis, ellipticis, planis; digitis posticis dimidio ferè palmatis, marginatis; dorso glandulis parvis instructo, fusco-griseo maculis, nigris, lineâ longitudinali pallidâ.

Habitat, Port Desire, Patagonia.

DESCRIPTION.—Head short. Muzzle rounded. Eyes prominent. Tongue thick, slightly heart-shaped, scarcely notched on the posterior margin. Palatine teeth prominent, in two oval groups, converging backwards. Tympanum rather small, perfectly round, conspicuous. Parotid glands distinct. Body thick and broad, with small glandular tubercles dispersed over the surface, particularly at the anterior part. Lumbar glands extremely large, elliptic-ovate, flat. Legs of moderate length, rather robust. Anterior toes separated, excepting at the base; a small tubercle under each joint, and several very small ones on the palm; hinder toes united to about half their length, and bordered on each side to the extremity; metatarsal tubercles prominent; soles of the hinder feet with many minute tubercles.

COLOUR.—The upper surface of this species is of a brownish grey colour, sometimes greenish brown or dark olive, and with numerous irregular spots of dark-brown or black. Thighs and legs with fasciæ of the same colour. Beneath yellowish white; in some with numerous blackish dots under the throat.

DIMENSIONS.

	In.	Lin.
Length of the head and body	1	8
of the anterior extremities	1	0
of the posterior extremities	2	3

Found by Mr. Darwin at Port Desire, in Patagonia, and high up the river Santa Cruz—"probably," says Mr. Darwin, "the most southern limit for this family."

GENUS—LEIUPERUS. *Bibr.*

LEIUPERUS SALARIUS. *Mihi.*

PLATE XVIII.—Fig. 1.

Supra nigricans, lumbis maculis 3 vel 4 nigris, albo-marginatis.

DESCRIPTION.—The head is short, the opening of the mouth small, the tongue rather thick, very slightly emarginate behind, and with the posterior margin free. The eyes small; the tympanum not very conspicuous; there is a trace of a parotid gland on each side of the neck. The body is rather thick, and the limbs proportionally short. The hinder toes are only connected at the base by a rudimentary membrane, the first four gradually increasing in length, and placed along the side of the matatarsus, one beyond the other; the fifth on the same line as the fourth, but not more than half its length. The metatarsal tubercle is rather prominent, and there are small subarticular tubercles on the toes of all the feet.

COLOUR.—The colour of the upper parts is brownish black. On each side near the thigh are three or four perfectly round black spots, each surrounded with a white line. The under parts whitish.

Of this second species of a rare and remarkable genus, one specimen only exists in Mr. Darwin's collection. It is only the third known instance, in the family of the RANIDÆ, of the absence of palatine teeth; the others being *Oxyglossus Lima* of Tschudi, and *Leiuperus marmoratus* of Bibron. The present genus must be considered as nearly approaching the family of the BUFONIDÆ in the absence of palatine and the extreme minuteness of the maxillary teeth, in the extremely small gape of the mouth, the thick form of the body, the shortness of the limbs, and the existence of rudimentary parotid glands. I have not had an opportunity of comparing this specimen with those on which Bibron founded the genus, but I cannot doubt the specific distinction between them.

DIMENSIONS.

	In.	Lin.
Length of the head and body	0	9
of the anterior extremities	0	5
of the posterior extremities	1	1

It was found by Mr. Darwin at Port Desire, and its habitat is very remarkable. "It is bred in and inhabits water far too salt to drink."

GENUS—PYXICEPHALUS. *Bibr.*

PYXICEPHALUS AMERICANUS. *Bibr.*

PLATE XVIII.—Fig. 2.

Linguá cordiformi; dentibus palatinis in lineá transversá interruptá, inter nares posteriores positis; tympano celato; dorso mammillato.

This curious species has, I believe, only once before been found. A single specimen exists in the French Museum, which was brought from Buenos Ayres by Mons. d'Orbigny, and which formed the subject of Mons. Bibron's description. Mr. Darwin's specimen was taken on the open plains at Monte Video.

Of the three species of this remarkable genus at present known, two are inhabitants of Africa, from whence they were brought by Delalande. As neither of them has as yet been figured, it was thought desirable that the present opportunity should be taken to exhibit some of the generic characters, and especially the hard horny spur on the hinder foot.

This genus is one of those bufonine forms of the RANIDÆ which irresistibly lead us to doubt the correctness of the present received arrangement of the anourous Amphibia.

Genus—ALSODES. *Bell.*

Caput *convexum.* Lingua *anticè acutè-producta, posticè rotundata, et libera.* Dentes palatini *inter nares posteriores.* .Tympanum *celatum.* Aperturæ Eustachianæ *haud conspicuæ.* Digiti anteriores *ad basin tantum*—posteriores *usque ad phalangem tertium membranâ connexi.*

A genus of the Raniform group, nearly allied, as Mons. Bibron observes, to *Scaphiopus,* by the structure of the hands, which, although without any projecting rudimentary thumb, has a small process under the skin, along the extreme margin of the first finger. In common with the genus *Bombinator,* it has the opening of the Eustachian tubes so small as scarcely to be detected.

Alsodes monticola. *Mihi.*

Plate XVIII.—Fig. 3.

Description.—Head semi-elliptical, somewhat convex, with the muzzle nearly perpendicular; vertex smooth. Eyes of moderate size. Nostrils very small, opening upwards. Tongue broad and rounded behind, narrowing to a point at the apex, detached at the posterior part. Palatine teeth in two small approximate patches, between the posterior nostrils. Openings of the Eustachian tubes scarcely visible. Extremities of moderate length. The fore feet, with four rather short toes, connected at the base by a short membrane; the inner toe broad, and with a slight projection under the skin, along its inner margin; hinder toes connected as far as the joint of the second and third phalanges.

The colour of the only specimen in the collection has become totally changed into a smoky brown by the spirit, but the following is Mr. Darwin's description

G

of it when living: " On the centre of the back a strong tinge of grass-green, shading on the sides into a yellowish brown ; iris coppery."

DIMENSIONS.

	In.	Lin.
Length of the head	0	5
of the body	1	0
of the anterior extremities	0	9
of the posterior extremities	1	8

Mr. Darwin found this species " in the island of Inchy, archipelago of Chonos, north part of Cape Tres Montes, from the same great height as *Bufo Chilensis* (from 500 to 2500 feet elevation) under a stone."

Genus—LITORIA. *Bibr.*

Litoria glandulosa. *Mihi.*

Plate XVIII.—Fig. 4.

Femoribus posticè glandulosis ; digitis posticis brevitèr palmatis.

This species agrees in many respects with *Litoria Americana* of Bibron. It differs, however, in the toes being much less palmate, at least according to the generic character given by that excellent naturalist, and in the existence of numerous thick glands on the posterior part of the thighs. The very slight degree to which the extremities of the toes are dilated in the other species of this genus, and which would at first sight lead to their allocation amongst the Rani-form rather than the Hyliform group, is in the present species even more strongly exhibited ; and it can scarcely be said that any dilatation exists at all.

The colours in the only specimen brought by Mr. Darwin are much obscured. The upper parts are apparently of an uniform brown, the under parts whitish, dotted with brown.

It was taken by Mr. Darwin at Concepcion, in Chile.

Genus—BATRACHYLA. *Bell.*

Lingua *suborbicularis, posticè libera.* Dentes palatini *in fasciculis binis obliquis inter nares posteriores dispositi.* Tympanum *distinctum, parvum, rotundum.* Digiti *depressi, ad apicem paullò dilatati, truncati.* Anteriores *ad basin tantum* —posteriores *paulò plus palmati.*

This genus, which considerably resembles *Hylodes,* is nevertheless sufficiently distinct from it, in the distribution of the palatine teeth, in the form of the dilatations of the toes, in the presence of a small palmar membrane, and some other points. One of the most remarkable of its characters is the form of the dilatation at the extremity of the toes; it is very small, transverse, truncated, and even a little emarginate; in this respect it must be considered as constituting a very near approach to the family of the RANIDÆ. We are unfortunately without any information as to the habits of the only known species which could throw any light upon its relations; but it is very clear that the dilatations of the toes are not such as to constitute it a true *tree-frog,* nor, on the other hand, are the connecting membranes of sufficient extent to give it the typical character of the swimming group of these animals.

BATRACHYLA LEPTOPUS. *Mihi.*

PLATE XVIII.—Fig. 5.

DESCRIPTION.—Head depressed, broad, rounded. Nostrils small, placed near together. Eyes large, opening considerably upwards. Tongue nearly round, the posterior part free for about one-third of its length. Palatine teeth placed in two small oval groups, placed obliquely, between the posterior nostrils, separated from each other by a considerable space. Tympanum small, nearly round. Limbs of moderate length. The toes on all the feet depressed, slender, the terminal dilatation very small, transverse, truncated; those of the fore feet connected at the base only, those of the hinder to the union of the first and second phalanges; of those of the fore feet the third is the longest, then the fourth, the second, and the first; of the hinder the fourth is the longest, then the third and fifth equal, then the second and the first. There are some minute scattered glands on the posterior part of the thighs.

The only specimen in Mr. Darwin's collection is in so bad a condition, that it is impossible to say with any certainty what is its natural colour. It is brown above, with a lighter band across the head between the eyes, and there are traces of a longitudinal line down the back ; the limbs are banded with brown and brownish yellow ; the under parts are pale, dotted with brown.

DIMENSIONS.

	In.	Lin.
Length of the head	0	5
of the body	1	0
of the anterior extremities	1	0
of the posterior extremities	2	1

Found by Mr. Darwin at Valdivia.

Genus—HYLORINA. *Bell.*

Caput *subrotundum planum.* Linguæ *magna circularis, posticè libera.* Dentes palatini *in lineâ transversâ, parum interruptâ, dispositi.* Tympanum *distinctum.* Digiti *subdepressi, ad apicem obtusi, haud expansi;* anteriores *ferè liberi;* posteriores *ad basin membranâ connexi, et marginati.* Femora *multò glandulosa.*

A genus nearly allied to *Hylodes*, from which, however, it may at once be distinguished by the palmure of the hinder toes—which in *Hylodes* are entirely free—and by the absence of even the slightest dilatation of their extremities ; offering another example of an osculant form between the HYLIDÆ and the RANIDÆ.

HYLORINA SYLVATICA. *Mihi.*

PLATE XIX.—FIG. 1.

DESCRIPTION.—Head broad, rounded, the anterior margin, from the nose to the lip, nearly perpendicular. Eyes large and prominent. Tympanum distinct, small, round. Tongue very large, circular, and entire, the posterior half free. Palatine teeth placed in a transverse line between the posterior nostrils, scarcely interrupted in the middle. Skin of the back rugose. Anterior feet with the toes long, rather slender, united at the base only by a very short membrane, with round subarticular tubercles, the apex rounded, but not presenting the slightest expansion. Hinder toes similarly formed, but with the connecting membrane more conspicuous, and extending along the sides of the toes nearly to the extremity. Thighs covered on the under and posterior surface with rather large and distinct glands.

The following is the description of the colouring, as given by Mr. Darwin from the living specimen. "Above fine grass green, mottled all over with copper colour, which nearly forms two longitudinal bands; beneath entirely of a lurid reddish lead colour. Iris brown."

DIMENSIONS.

	In.	Lin.
Length of the head	1	0
of the body	1	8
of the anterior extremities	2	1
of the posterior extremities	4	4

Found by Mr. Darwin in the Archipelago of Chonos (S. of Chiloe) in thick forests.

GENUS—HYLA.

HYLA VAUTERII. *Bibr*.

PLATE XIX. FIG. 2.

Linguá subcordiformi, posticè emarginatá. Dentibus palatinis in fasciculis binis ovalilibus, subcontiguis. Oculis prominentibus. Capite tam lato quam longo. Gulá bi-plicatá; suprà levitèr—infrà multùm granulosa. Dorso fusco-griseo, punctis, maculis et fasciis lateralibus nigris.

Hyla Vauterii. Bibr. MS.

DESCRIPTION.—Head short, thick, the sides anteriorly converging towards a nearly right angle, the muzzle rounded. Tongue nearly cordate, posteriorly emarginate, free for about one-fourth of its length. Palatine teeth in two oval fasciculi, placed nearly transversely between the posterior nares, and almost contiguous. Eyes prominent. Tympanum circular, rather large. Body plump, the sides nearly parallel for two-thirds of its length. The skin nearly smooth, but covered with very small inconspicuous granulations over the whole upper surface, which are rather more obvious on the head. The throat, the belly, and the inferior surface of the thighs covered with large prominent granulations. Beneath the lower jaw the granulations are smaller, and the under surface of the limbs excepting the thighs is quite smooth. A small fold of skin over the tympanum passes backwards to the arm; and beneath the throat there are two considerable transverse folds, one of which is before and the other immediately behind the arms. Fore feet, with the palms covered with small granular tubercles, and a tubercle under the joints of the fingers, which are connected to about one-third of their length. Hinder legs longer than the head and body by the whole foot and tarsus. The soles tuberculated. Toes rather short, palmate to half their length.

COLOUR.—The whole of the upper parts are greyish brown, with a tinge of red, and minutely punctured with black. There are scattered spots of the latter colour on the back and sides, assuming somewhat of a longitudinal arrangement, and a broad blackish grey fascia extends from the eye backwards to the arm, including the tympanum, and this fascia is bordered beneath by a white line. The thighs and legs are barred and spotted with black. The under parts are yellowish white, excepting under the lower jaw, where it is finely mottled with black and white.

I received the name of this species from Mons. Bibron, who had, I believe, applied it to specimens in the Paris Museum. It was taken by Mr. Darwin at Maldonado, lurking under a stone, and at Rio Janeiro on palm-trees.

DIMENSIONS.

	In.	Lin.
Length of the head and body	1	6
of anterior extremities	0	9
of posterior extremities	2	5

HYLA AGRESTIS. *Mihi.*

PLATE XIX.—FIG. 3.

Capite brevi. Oculis subprominentibus. Tympano mediocri circulari. Linguâ sub-rotundâ, posticè liberâ, anticè angustatâ. Dentibus palatinis in fasciculis binis, paulò separatis, ad marginem postico-interiorem narium posteriorum. Dorso granu-loso. Gulâ plicatâ. Digitis anticis ad basin tantùm, posticis usque ad phalanges penultimas palmatis. Suprà viridis, linea albâ laterali, femoribus posticè atque lateribus abdominis, albis, nigro-maculatis.

Habitat, Maldonado, in grassy fields.

DESCRIPTION.—Head short, thick, the two sides of the muzzle approaching each other at a rather acute angle, rounded at the extremity. Eyes rather large and prominent. Tympanum circular, of moderate size, and very distinct. Tongue entire, rounded, and free behind, narrowed, and almost angular in front. Palatine teeth in two oval parcels separated by a very small interval, and placed on a line with the hinder margin of the *posterior nares;* the whole of the back covered with extremely small granules; a slight fold or elevation of the skin commencing above the posterior margin of the tympanum, and extending backwards just above the arm, in front of which it is met by a more considerable one which crosses the throat from side to side; the under parts covered with large granules; fore feet with the toes palmated only at the base; hinder ones palmated to four-fifths of the length.

The colour of this beautiful species is thus given by Mr. Darwin :—" Above emerald green, beneath white; a silvery white stripe bordered beneath with a very narrow black line, extends from the corner of the eye, along the side, to the thigh; a smaller one at the corner of the mouth; the posterior surface of the hinder legs and the flanks marked with black spots. Iris gold coloured; tympanum brown."

DIMENSIONS.

	In.	Lin.
Length of the head and body	1	8
of the anterior extremities......................	1	0
of the posterior extremities	2	8

The young of this species, instead of the bright green colour of the upper parts, is of a delicate grey with small brown markings; and a lateral fascia of brown, bordered above and beneath with a white line, extends from the fore part of the head backwards, the upper white line nearly to the thigh, the inferior one to the shoulder. The black spots on the flanks and thighs are but just visible.

This species so nearly resembles the *Hyla pulchella* of Mons: Bibron, at least as far as his description enables me to ascertain its characters, that it was with some hesitation that I came to the conclusion that they are distinct. Exclusive, however, of the difference of colour, the back of the present species is granulated, and the throat still more distinctly so, whereas the other animal has the skin on the upper parts, as well as on the anterior part of the throat, quite smooth. The palatine teeth also appear to be somewhat differently arranged.

Mr. Darwin observes, that this species was found in numbers in the open grass plains, and likewise in swamps, about Maldonado, and that they can never ascend trees, as these are entirely wanting at the places frequented by the Hylæ.

Fam.—BUFONIDÆ.

Genus—RHINODERMA. Bibr.

Lingua *cordato-ovata, postice libera et subemarginata.* Dentes palatini *nulli.* Tympanum *celatum.* Glandæ parotideæ *nullæ.* Digiti *breves, depressi;* anteriores *ad basin tantùm,* posteriores *ferè dimidio palmati.* Rostrum *cutis appendiculo filiformi instructum.*

RHINODERMA DARWINII. *Bibr.*

PLATE XX.—FIG. 1, 2.

Suprà pallidè rufo-cinereum, fasciis transversis viridescentibus ; subtus castaneo-nigrum, maculis albis.

Rhinoderma Darwinii. Bibr. Hist. Nat. Rept. VIII. p. 659. *Var.* Dorso fuscescenti-nigro.

DESCRIPTION.—The head and body are flattened, the head triangular, slightly truncated in front, but appearing angular from the skin being produced into a small filiform appendage, standing forwards from the extremity of the snout. The eyes are lateral, slightly prominent. Body very slender. Skin perfectly smooth, and without apparent glands, excepting on the thighs. Fore legs rather short, reaching quite to the thighs when placed by the side; the toes almost wholly separate, there being but the rudiment of a connecting membrane at their base. Hinder legs long, extending forwards beyond the head by the whole length of the foot; the hinder toes are connected nearly half their length, and the connecting membrane is thick and coloured like the rest of the skin.

COLOUR.—The colour varies greatly in different individuals. The following are the principal variations in the specimens collected by Mr. Darwin. Above pale iron rust-colour, with a transverse fascia across the head, a triangular one over the shoulders, a large broad mark on the loins, and the upper part of the thighs all of a bright beautiful green. The under side anteriorly rich chestnut-brown, passing into black posteriorly, with several irregular snow-white spots, particularly a broad one across the belly, and white bands across the legs. Another specimen was cream colour above, the markings darker, and with small spots of green. In one the chestnut colour beneath was replaced by bright yellow. There is one, constituting a very distinct variety, in which the upper part is wholly and almost uniformly dark brown. The female is greenish grey above, without conspicuous markings.

This is the only known species of the genus, which was founded by Mons. Bibron upon the specimens collected by Mr. Darwin. The general slightness and elegance of its form, and its slender proportions, would lead us to consider it at first sight as rather belonging to the *Ranidæ* than the *Bufonidæ ;* but the total absence of teeth in the upper maxillary arch, shews that its proper place is in the latter group. Its form and the length of the posterior extremities would also prepare us to expect that it can leap freely, which Mr. Darwin states to be the fact. It inhabits thick and gloomy forests, and is excessively common in the forest of Valdivia.

DIMENSIONS.

	In.	Lin.
Length of the head and body	1	0
of the anterior extremities	0	5
of the posterior extremities	1	4

Bufo Chilensis. *Bibr.*

Of this species, which has been described under different names by many naturalists, and the synonymy of which has only lately been cleared up by Mons. Bibron, there exist numerous specimens in the collection of Mr. Darwin, who found it at Buenos Ayres, and also in the Archipelago of Chonos, on the west coast of South America. It is certainly remarkable that the same species should be found on the opposite sides of the Continent; but on a careful examination I do not find any specific distinctions between the specimens from the different localities. The Prince de Wied has described it as found at Brazil, under the name of *Bufo cinctus*, and it is also well known as having been repeatedly procured in Peru and in Chile; but Mons. Bibron has in his work considered them all as belonging to but one species. The following account of its habits as given by Mr. Darwin is very curious and interesting :— " These Toads are exceedingly abundant all over the treeless damp mountains of granite, crawling about, and eating during the daytime, and making a noise similar to that which is commonly used in England to quicken horses. Many of them on being touched close their eyes, arch their back, and draw up their legs (as if the spinal marrow was divided), probably as an artifice. They are remarkable from their curious manner of *running* like the Natter Jack of England ; they scarcely ever jump, neither do they crawl like a toad, but run very quickly. Their bright colours give them a very strange appearance. They abound at an elevation of 500 to 2500 feet."

Genus—PHRYNISCUS.

Phryniscus nigricans. *Weigm.*

Plate XX.—Fig. 3, 4, 5.

Dorso granuloso, scabriusculo. Pedibus posticis subpalmatis. Corpore membrisque nigris, abdomine maculâ magnâ transversâ ad partem posteriorem et maculâ rotunda utrinque medium versus, palmis atque plantis, omnibus coccineis.

Phryniscus nigricans. Weigm. Nov. Act. Leop. XVII. p. 264. Bibr. Hist. Rept. VIII. p. 723.
Chaunus formosus. Tschudi Classif. Batrach.

Habitat, Maldonado and Bahia Blanca.

H

This curious little species has been described by Weigman under the present name,—by Tschudi under the generic name of *Chaunus*, and fully by Bibron, who retained the name originally given to it by Weigman. It now remains only to correct, from Mr. Darwin's notes, some points respecting the colours, which had been mis-stated in consequence of the action of the spirit in which the specimen had been preserved. The colour of this curious miniature representation of a Toad, is "ink black," excepting the palms and soles of the feet, a large transverse spot across the posterior part of the abdomen, two smaller ones near the middle, and in some specimens a few scattered little spots, all of the most intense vermilion red. There is one specimen from Bahia Blanca which has also some small "buff-orange" spots on the upper part.* Mr. Darwin observes that "the appearance of the vermilion colour is as if the animal had crawled over a newly painted board;" and he adds—"This Toad inhabits the most dry and sandy plains of Bahia Blanca, where there is no appearance of water ever lodging." The other specimens were taken at Maldonado, where it inhabits the sand-dunes near the coast. Mr. Darwin threw one into a pool of fresh-water, but he found it could hardly swim, and he thinks, if unassisted, it would have been soon drowned.

This species is diurnal in its habits, and may be daily seen under a scorching sun, crawling over the parched and loose sand. M. D'Orbigny brought specimens from Monte Video.

DIMENSIONS.

	In.	Lin.
Length of the head and body	1	0
of the anterior extremities	0	5
of the posterior extremities..............	0	8

Genus—UPERODON. *Bibr.*

UPERODON ORNATUM. *Mihi.*

PLATE XX.—Fig. 6.

Capite multò latiore quam longiore.   Dorso olivaceo, maculis fuscis, albo marginatis.

Habitat Buenos Ayres.

DESCRIPTION.—Head more than half as broad again as it is long, and equal in breadth to half the entire length of the head and body. Muzzle rounded. Nostrils oval, opening upwards and a

---

* This specimen from Bahia Blanca has a much smoother skin than the others; but from its similarity in all other characters there can be no doubt of its specific identity with them.

little outwards. Eyes rather large, the upper eyelids forming perfect flaps, which entirely cover the eyes. Body rounded, very broad. The shoulders and thighs wholly concealed by the skin of the body. Limbs very short. The anterior feet very broad. The toes somewhat depressed, very short, bordered with a fold of skin. Hinder feet with the toes more depressed and more distinctly bordered. Back covered with small glands.

COLOUR.—The colour of the upper surface is dark olive, becoming lighter at the sides, and having numerous dark brown spots, which are round, oval, elliptical, or irregular, of very various sizes, placed somewhat symmetrically, and each bordered with a whitish or yellow line. Beneath pale, excepting the throat, which is black.

I have ventured to consider this remarkable amphibian as specifically distinct from *U. marmoratum* of Bibron ; a conclusion to which I have been almost imperatively led, by the fact of its inhabiting a different hemisphere from all known specimens of that species. The other was found by M. Leschenault in the interior of the peninsula of India : the specimen from which the present description is taken was obtained by Mr. Darwin at Buenos Ayres. Notwithstanding the similarity of the two species, which is so great as to have led Mons. Bibron to consider them as identical, I could not assent to such an anomaly as the existence of an animal, at once so rare and possessed of such limited powers of locomotion, in two regions so widely remote. I have not the opportunity of comparing the specimens of the former species with the present, but, even from Mons. Bibron's description, I believe that I can discover sufficient discrepancies between the animals, to bear me out in the view I have taken. These discrepancies I venture to place in the following tabular view, and leave zoologists to form their own conclusions.

UPERODON MARMORATUM.

" La tête offre en arrière une largeur à peu près égale à son longueur totale, laquelle entre pour le quart environ dans l'étendue de l'animal."

" On pourrait considérer la peau comme étant parfaitement lisse, si l'on ne voyait éparses sur le dessus du tronc un certain nombre de verrues glanduleuses d'un assez grand diamètre relativement à la grosseur de l'animal, mais fort peu saillantes ou à peine convexes."

" Les parties supérieures de ce Batracien présentent sur un fond olivâtre, d'énormes tâches brunes, *toutes confluentes, ou s'anastomosant diversement.*" *

UPERODON ORNATUM.

Head fully half as broad again as it is long, and equal in breadth to half the total length of the animal.

Back covered with numerous *small* glandular tubercles, notably elevated.

All the spots on the back are quite distinct, not in any way passing into each other or connected, and each encircled by a white line.

* Bibr. Rept. VIII. p. 749.

Plate 1

1. *Proctotretus Chilensis.*
2. ——————— *gracilis.* } *Nat size*
1a 1b.} *Magnified parts.*
2a.}

Drawn from Nature by ... Waterhouse. Faciunt
on stone in Lithotint. C.Hullmandel's Patent

1 2 *Proctotretus pulcher*. Var. ...

Plate 3.

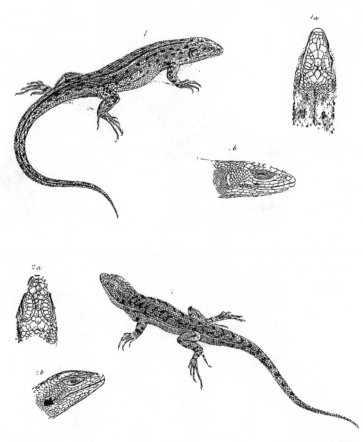

Nature by B. Waterhouse Hawkins
Printed C. Hullmandel's Patent

1. *Proctotretus Bibronii* ⎫ *Nat. Size.*
2. _____ *tenuis* ⎭
a & b. 1 & 2. *Magnified Views of Heads.*

Plate 4.

1. *Proctotretus signifer.*
2. _____ *nigromaculatus.*
2 a _____ *Magnified View.*

Plate 5.

Drawn by B. Waterhouse Hawkins.
Lithod. C. Hullmandel's Patent.

1 *Proctotretus Fitzingerii*
2 ————— *Cyanogaster*
        Nat. Size

Plate 6.

a

b

Proctotretus tenuis

Plate 7.

From Nature by B. Waterhouse Hawkins.
in Litho at C. Hullmandels Patent

1
2 } Proctotretus Darwinii. Nat size.

1 a & b
2 a. } Magnified Views.

Plate 8.

B. Waterhouse Hawkins lithog

1 }
2 } *Proctotretus Weigmannii.*

1a }
2a } *Magnified Views.*

Plate 9.

1a.

1b.

2a.

Drawn by B. Waterhouse Hawkins,
Printed C. Hullmandel's Patent.

1. *Tropidurus multimaculatus* } *Nat. size.*
2. _____ *pectinatus.*

1a. 1b } *Magnified Views.*
2a.

Diplolæmus Darwinii Nat. size.

Plate 11

Diplolaemus Bibronii var. ...

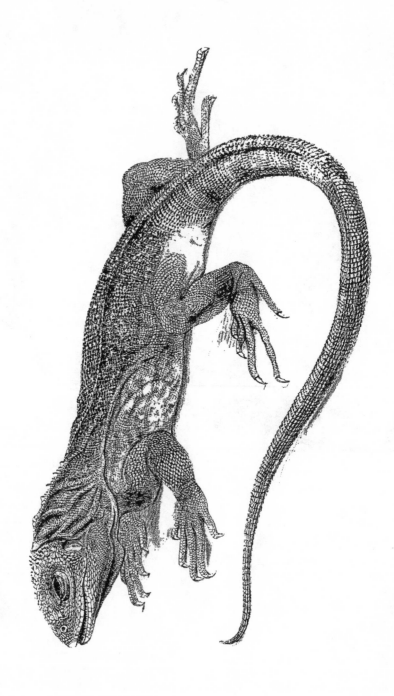

1. *Gymnodactylus Gaudichaudii.*
2. *Naultinus Grayii.*

1. Leiocephalus Gruyi
2. Centrura flagellifer.

1.   Ameiva longicauda.
2. 2a. Ptycholoemus gemmatus } ...
3.   Cyclodus Carinatus)

1. Rana Delalandii.
2. Rana Mascariensis.
3. 3a. Limnodytes punctus.   } Nat. size.
4. Cystignathus bergianus.

1. *Borborocœtes Bibronii.* 1a. Mag. View of Tongue & Gullet.
2.   *Grayii.*
3. *Pleurodema Darwinii.* } Nat. Size
4.   *elegans.*
5.   *bufoninum.*

1. 1a.   *Leiuperus salarius.*
2. 2a. 2b. 2c. *Pyxicephalus Americanus.*
3. 3a. 3b.   *Alsodes monticola.*
4. 4a.   *Litoria glandulosa.*
5. 5a. 5b.   *Batrachyla leptopus.*

1. 1a. *Hylania cyclacans*
2. 2a. *Hyla aquarius*

1. 2. *Rhinoderma Darwini.*
3. 4. 5. *Pleurodema nigricans.*
3. *Cystignathus ornatum.*